ECONOMIC ISSUES, PROBLEMS AND PERSPECTIVES

OFFSHORE PROFIT SHIFTING AND U.S. TAX CODE WEAKNESSES

ANALYSES AND THE APPLE INC. CASE STUDY

ECONOMIC ISSUES, PROBLEMS AND PERSPECTIVES

Additional books in this series can be found on Nova's website
under the Series tab.

Additional E-books in this series can be found on Nova's website
under the E-book tab.

ECONOMIC ISSUES, PROBLEMS AND PERSPECTIVES

OFFSHORE PROFIT SHIFTING AND U.S. TAX CODE WEAKNESSES

ANALYSES AND THE APPLE INC. CASE STUDY

RENY TOUPIN
EDITOR

nova
publishers
New York

Copyright © 2013 by Nova Science Publishers, Inc.

For permission to use material from this book please contact us:
Telephone 631-231-7269; Fax 631-231-8175
Web Site: http://www.novapublishers.com

NOTICE TO THE READER

The Publisher has taken reasonable care in the preparation of this book, but makes no expressed or implied warranty of any kind and assumes no responsibility for any errors or omissions. No liability is assumed for incidental or consequential damages in connection with or arising out of information contained in this book. The Publisher shall not be liable for any special, consequential, or exemplary damages resulting, in whole or in part, from the readers' use of, or reliance upon, this material. Any parts of this book based on government reports are so indicated and copyright is claimed for those parts to the extent applicable to compilations of such works.

Independent verification should be sought for any data, advice or recommendations contained in this book. In addition, no responsibility is assumed by the publisher for any injury and/or damage to persons or property arising from any methods, products, instructions, ideas or otherwise contained in this publication.

This publication is designed to provide accurate and authoritative information with regard to the subject matter covered herein. It is sold with the clear understanding that the Publisher is not engaged in rendering legal or any other professional services. If legal or any other expert assistance is required, the services of a competent person should be sought. FROM A DECLARATION OF PARTICIPANTS JOINTLY ADOPTED BY A COMMITTEE OF THE AMERICAN BAR ASSOCIATION AND A COMMITTEE OF PUBLISHERS.

Additional color graphics may be available in the e-book version of this book.

Library of Congress Cataloging-in-Publication Data

ISBN: 978-1-62808-479-5

Published by Nova Science Publishers, Inc. † New York

CONTENTS

PREFACE

This book examines how Apple Inc. transferred the economic rights to its intellectual property through a cost sharing agreement with its own offshore affiliates, and was thereby able to shift tens of billions of dollars offshore to a low tax jurisdiction and avoid U.S. tax. Apple Inc then utilized U.S. tax loopholes, including the so-called "check-the-box" rules, to avoid U.S. taxes on $44 billion in taxable offshore income over the past four years, or about $10 billion in tax avoidance per year. The book also examines some of the weaknesses and loopholes in certain U.S. tax code provisions, including transfer pricing, Subpart F, and related regulations, that enable multinational corporations to avoid U.S. taxes.

Chapter 1 – On May 21, 2013, the Permanent Subcommittee on Investigations (PSI) of the U.S. Senate Homeland Security and Government Affairs Committee held a hearing that is a continuation of a series of reviews conducted by the Subcommittee on how individual and corporate taxpayers are shifting billions of dollars offshore to avoid U.S. taxes. The hearing examined how Apple Inc., a U.S. multinational corporation, has used a variety of offshore structures, arrangements, and transactions to shift billions of dollars in profits away from the United States and into Ireland, where Apple has negotiated a special corporate tax rate of less than two percent. One of Apple's more unusual tactics has been to establish and direct substantial funds to offshore entities in Ireland, while claiming they are not tax residents of any jurisdiction. For example, Apple Inc. established an offshore subsidiary, Apple Operations International, which from 2009 to 2012 reported net income of $30 billion, but declined to declare any tax residence, filed no corporate income tax return, and paid no corporate income taxes to any national government for five years. A second Irish affiliate, Apple Sales International, received $74 billion in sales income over four years, but due in part to its alleged status as a non-tax resident, paid taxes on only a tiny fraction of that income. In addition, the hearing examined how Apple Inc. transferred the economic rights to its intellectual property through a cost sharing agreement with its own offshore affiliates, and was thereby able to shift tens of billions of dollars offshore to a low tax jurisdiction and avoid U.S. tax. Apple Inc. then utilized U.S. tax loopholes, including the so-called "check-the-box" rules, to avoid U.S. taxes on $44 billion in taxable offshore income over the past four years, or about $10 billion in tax avoidance per year. The hearing also examined some of the weaknesses and loopholes in certain U.S. tax code provisions, including transfer pricing, Subpart F, and related regulations, that enable multinational corporations to avoid U.S. taxes.

Chapter 2 – This is the Testimony of Apple Inc. Hearing on "Offshore Profit Shifting and the U.S. Tax Code - Part 2 (Apple Inc.)."

Chapter 3 – This is the Testimony of J. Richard Harvey, Jr., Distinguished Professor, Villanova University School of Law. Hearing on "Offshore Profit Shifting and the U.S. Tax Code - Part 2 (Apple Inc.)."

Chapter 4 – This is the Testimony of Stephen E. Shay, Professor of Practice, Harvard Law School. Hearing on "Offshore Profit Shifting and the U.S. Tax Code - Part 2 (Apple Inc.)."

Chapter 5 – This is the Testimony of Mark J. Mazur, Assistant Secretary for Tax Policy, U.S. Department of the Treasury. Hearing on "Offshore Profit Shifting and the U.S. Tax Code - Part 2 (Apple Inc.)."

Chapter 6 – This is the Testimony of Samuel M. Maruca, Director, Transfer Pricing Operations, Internal Revenue Service. Hearing on "Offshore Profit Shifting and the U.S. Tax Code - Part 2 (Apple Inc.)."

In: Offshore Profit Shifting ...   ISBN: 978-1-62808-479-5
Editor: Reny Toupin   © 2013 Nova Science Publishers, Inc.

Chapter 1

OFFSHORE PROFIT SHIFTING AND THE U.S. TAX CODE - PART 2 (APPLE INC.)*

Permanent Subcommittee on Investigations

MEMORANDUM

TO: Members of the Permanent Subcommittee on Investigations
FROM: Senator Carl Levin, Chairman
           Senator John McCain, Ranking Minority Member Permanent Subcommittee on Investigations
DATE: May 21, 2013
RE: Offshore Profit Shifting and the U.S. Tax Code – Part 2 (Apple Inc.)

I. EXECUTIVE SUMMARY

On May 21, 2013, the Permanent Subcommittee on Investigations (PSI) of the U.S. Senate Homeland Security and Government Affairs Committee held a hearing that is a continuation of a series of reviews conducted by the Subcommittee on how individual and corporate taxpayers are shifting billions of dollars offshore to avoid U.S. taxes. The hearing examined how Apple Inc.,

---

* This is an edited, reformatted and augmented version of Permanent Subcommittee on Investigations, dated May 21, 2013.

a U.S. multinational corporation, has used a variety of offshore structures, arrangements, and transactions to shift billions of dollars in profits away from the United States and into Ireland, where Apple has negotiated a special corporate tax rate of less than two percent. One of Apple's more unusual tactics has been to establish and direct substantial funds to offshore entities in Ireland, while claiming they are not tax residents of any jurisdiction. For example, Apple Inc. established an offshore subsidiary, Apple Operations International, which from 2009 to 2012 reported net income of $30 billion, but declined to declare any tax residence, filed no corporate income tax return, and paid no corporate income taxes to any national government for five years. A second Irish affiliate, Apple Sales International, received $74 billion in sales income over four years, but due in part to its alleged status as a non-tax resident, paid taxes on only a tiny fraction of that income.

In addition, the hearing examined how Apple Inc. transferred the economic rights to its intellectual property through a cost sharing agreement with its own offshore affiliates, and was thereby able to shift tens of billions of dollars offshore to a low tax jurisdiction and avoid U.S. tax. Apple Inc. then utilized U.S. tax loopholes, including the so-called "check-the-box" rules, to avoid U.S. taxes on $44 billion in taxable offshore income over the past four years, or about $10 billion in tax avoidance per year. The hearing also examined some of the weaknesses and loopholes in certain U.S. tax code provisions, including transfer pricing, Subpart F, and related regulations, that enable multinational corporations to avoid U.S. taxes.

A. Subcommittee Investigation

For a number of years, the Subcommittee has reviewed how U.S. citizens and multinational corporations have exploited and, at times, abused or violated U.S. tax statutes, regulations and accounting rules to shift profits and valuable assets offshore to avoid U.S. taxes. The Subcommittee inquiries have resulted in a series of hearings and reports.[1] The Subcommittee's recent reviews have focused on how multinational corporations have employed various complex structures and transactions to exploit taxloopholes to shift large portions of their profits offshore and dodge U.S. taxes.

At the same time as the U.S. federal debt has continued to grow – now surpassing $16 trillion – the U.S. corporate tax base has continued to decline, placing a greater burden on individual taxpayers and future generations. According to a report prepared for Congress:

"At its post-WWII peak in 1952, the corporate tax generated 32.1% of all federal tax revenue. In that same year the individual tax accounted for 42.2% of federal revenue, and the payroll tax accounted for 9.7% of revenue. Today, the corporate tax accounts for 8.9% of federal tax revenue, whereas the individual and payroll taxes generate 41.5% and 40.0%, respectively, of federal revenue."[2]

Over the past several years, the amount of permanently reinvested foreign earnings reported by U.S. multinationals on their financial statements has increased dramatically. One study has calculated that undistributed foreign earnings for companies in the S&P 500 have increased by more than 400%.[3] According to recent analysis by Audit Analytics, over a five year period from 2008 to 2012, total untaxed indefinitely reinvested earnings reported in 10-K filings for firms comprising the Russell 3000 increased by 70.3%.[4] During the same period, the number of firms reporting indefinitely reinvested earnings increased by 11.4%.

The increase in multinational corporate claims regarding permanently reinvested foreign earnings and the decline in corporate tax revenue are due in part to the shifting of mobile income offshore into tax havens. A number of studies show that multinational corporations are moving "mobile" income out of the United States into low or no tax jurisdictions, including tax havens such as Ireland, Bermuda, and the Cayman Islands.[5] In one 2012 study, a leading expert in the Office of Tax Analysis of the U.S. Department of Treasury found that foreign profit margins, not foreign sales, are the cause for significant increases in profits abroad. He wrote:

"The foreign share of the worldwide income of U.S. multinational corporations (MNCs) has risen sharply in recent years. Data from a panel of 754 large MNCs indicate that the MNC foreign income share increased by 14 percentage points from 1996 to 2004. The differential between a company's U.S. and foreign effective tax rates exerts a significant effect on the share of its income abroad, largely through changes in foreign and domestic profit margins rather than a shift in sales. U.S.-foreign tax differentials are estimated to have raised the foreign share of MNC worldwide income by about 12 percentage points by 2004. Lower foreign effective tax rates had no significant effect on a company's domestic sales or on the growth of its worldwide pre-tax profits. Lower taxes on foreign income do not seem to promote 'competitiveness.'"[6]

One study showed that foreign profits of controlled foreign corporations (CFCs) of U.S. multinationals significantly outpace the total GDP of some tax havens."[7] For example, profits of CFCs in Bermuda were 645% and in the Cayman Islands were 546% as a percentage of GDP, respectively. In a recent research report, JPMorgan expressed the opinion that the transfer pricing of intellectual property "explains some of the phenomenon as to why the balances of foreign cash and foreign earnings at multinational companies continue to grow at such impressive rates." [8] On September 20, 2012, the Subcommittee held a hearing and examined some of the weaknesses and loopholes in certain tax and accounting rules that facilitated profit shifting by multinational corporations. Specifically, it reviewed transfer pricing, deferral, and Subpart F of the Internal Revenue Code, with related regulations, and accounting standards governing offshore profits and the reporting of tax liabilities. The Subcommittee presented two case studies: (1) a study of structures and practices employed by Microsoft Corporation to shift and keep profits offshore; and (2) a study of Hewlett-Packard's "staggered foreign loan program," which was devised to *de facto* repatriate offshore profits to the United States to help run its U.S. operations, without paying U.S. taxes. The case study for the Subcommittee's May 2013 hearing involves Apple Inc. Building upon information collected in previous inquiries, the Subcommittee reviewed Apple responses to several Subcommittee surveys, reviewed Apple SEC filings and other documents, requested information from Apple, and interviewed a number of corporate representatives from Apple. The Subcommittee also consulted with a number of tax experts and the IRS. This memorandum first provides an overview of certain tax provisions related to offshore income, such as transfer pricing, Subpart F, and the so-called check-the-box regulations and look-through rule. It then presents a case study of Apple's organizational structure and the provisions of the tax code and regulations it uses to shift and keep billions in profits offshore in two controlled foreign corporations formed in Ireland. The first is Apple Sales International (ASI), an entity that has acquired certain economic rights to Apple's intellectual property. Apple Inc. has used those rights of ASI to shift billions in profits away from the United States to Ireland, where it pays a corporate tax rate of 2% or less. The second is Apple Operations International (AOI), a 30-year old corporation that has no employees or physical presence, and whose operations are managed and controlled out of the United States. Despite receiving $30 billion in earnings and profits during the period 2009 through 2011 as the key holding company for Apple's extensive offshore corporate structure, Apple Operations International has no declared tax residency anywhere in the world and, as a consequence, has not paid corporate income tax to any national government for the past 5 years. Apple has recently disclosed that ASI also claims to have no tax residency in any jurisdiction, despite receiving over a four year period from 2009 to 2012, sales income from Apple affiliates totaling $74 billion. Apple is an American success story. Today, Apple Inc. maintains more than $102 billion in offshore cash, cash equivalents and marketable securities (cash).[9] Apple executives told the Subcommittee that the company has no intention of returning those funds to the United States unless and until there is a more favorable environment, emphasizing a lower corporate tax rate and a simplified tax code.[10] Recently, Apple issued $17 billion in debt instruments to provide funds for its U.S. operations rather than bring its offshore cash home, pay the tax owed, and use those funds to invest in its operations or return dividends to its stockholders. The Subcommittee's investigation shows that Apple has structured organizations and business operations to avoid U.S. taxes and reduce the contribution it makes to the U.S. treasury. Its actions disadvantage Apple's domestic competitors, force other taxpayers to shoulder the tax burden Apple has cast off, and undermine the fairness of the U.S. tax code. The purpose of the Subcommittee's investigation is to describe Apple's offshore tax activities and offer recommendations to close the offshore tax loopholes that enable some U.S. multinational corporations to avoid paying their share of taxes.

B. Findings and Recommendations

Findings

The Subcommittee's investigation has produced the following findings of fact.

1. **Shifting Profits Offshore.** Apple has $145 billion in cash, cash equivalents and marketable securities, of which $102 billion is "offshore." Apple has used offshore entities, arrangements, and transactions to transfer its assets and profits offshore and minimize its corporate tax liabilities.

2. **Offshore Entities With No Declared Tax Jurisdiction.** Apple has established and directed tens of billions of dollars to at least two Irish affiliates, while claiming neither is a tax resident of any jurisdiction, including its primary offshore holding company, Apple Operations International (AOI), and its primary intellectual property rights recipient, Apple Sales International (ASI). AOI, which has no employees, has no physical presence, is managed and controlled in the

United States, and received $30 billion of income between 2009 and 2012, has paid no corporate income tax to any national government for the past five years.

3. **Cost Sharing Agreement.** Apple's cost sharing agreement (CSA) with its offshore affiliates in Ireland is primarily a conduit for shifting billions of dollars in income from the United States to a low tax jurisdiction. From 2009 to 2012, the CSA facilitated the shift of $74 billion in worldwide sales income away from the United States to Ireland where Apple has negotiated a tax rate of less than 2%.

4. **Circumventing Subpart F.** The intent of Subpart F of the U.S. tax code is to prevent multinational corporations from shifting profits to tax havens to avoid U.S. tax. Apple has exploited weaknesses and loopholes in U.S. tax laws and regulations, particularly the "check-the-box" and "look-through" rules, to circumvent Subpart F taxation and, from 2009 to 2012, avoid $44 billion in taxes on otherwise taxable offshore income.

Recommendations

Based upon the Subcommittee's investigation, the Memorandum makes the following recommendations.

1. **Strengthen Section 482.** Strengthen Section 482 of the tax code governing transfer pricing to eliminate incentives for U.S. multinational corporations to transfer intellectual property to shell entities that perform minimal operations in tax haven or low tax jurisdictions by implementing more restrictive transfer pricing rules concerning intellectual property.

2. **Reform Check-the-Box and Look Through Rules.** Reform the "check-the-box" and "look-through" rules so that they do not undermine the intent of Subpart F of the Internal Revenue Code to currently tax certain offshore income.

3. **Tax CFCs Under U.S. Management and Control.** Use the current authority of the IRS to disregard sham entities and impose current U.S. tax on income earned by any controlled foreign corporation that is managed and controlled in the United States.

4. **Properly Enforce Same Country Exception.** Use the current authority of the IRS to restrict the "same country exception" so that the exception to Subpart F cannot be used to shield from taxation passive income shifted between two related entities which are

incorporated in the same country, but claim to be in different tax residences without a legitimate business reason.

5. **Properly Enforce the Manufacturing Exception.** Use the current authority of the IRS to restrict the "manufacturing exception" so that the exception to Subpart F cannot be used to shield offshore income from taxation unless substantial manufacturing activities are taking place in the jurisdiction where the intermediary CFC is located.

II. OVERVIEW OF TAX PRINCIPLES AND LAW

A. U.S. Worldwide Tax and Deferral

U.S. corporations are subject to a statutory tax rate of up to a 35% on all their income, including worldwide income, which on its face is a rate among the highest in the world. This statutory tax rate can be reduced, however, through a variety of mechanisms, including tax provisions that permit multinational corporations to defer U.S. tax on active business earnings of their CFCs until those earnings are brought back to the United States, *i.e.*, repatriated as a dividend. The ability of a U.S. firm to earn foreign income through a CFC without US tax until the CFC's earnings are paid as a dividend is known as "deferral." Deferral creates incentives for U.S. firms to shift U.S. earnings offshore to low tax or no tax jurisdictions to avoid U.S. taxes and increase their after tax profits. In other words, tax haven deferral is done for tax avoidance purposes.[11] U.S. multinational corporations shift large amounts of income to low-tax foreign jurisdictions, according to a 2010 report by the Joint Committee on Taxation.[12] Current estimates indicate that U.S. multinationals have more than $1.7 trillion in undistributed foreign earnings and keep at least 60% of their cash overseas.[13] In many instances, the shifted income is deposited in the names of CFCs in accounts in U.S. banks.[14] In 2012, President Barack Obama reiterated concerns about such profit shifting by U.S multinationals and called for this problem to be addressed through tax reform.[15]

B. Transfer Pricing

A major method used by multinationals to shift profits from high-tax to low-tax jurisdictions is through the pricing of certain intellectual property rights, goods and services sold between affiliates. This concept is known as "transfer pricing." Principles regarding transfer pricing are codified under Section 482 of the Internal Revenue Code and largely build upon the principle of arms length dealings. IRS regulations provide various economic methods that can be used to test the arm's length nature of transfers between related parties. There are several ways in which assets or services are transferred between a U.S. parent and an offshore affiliate entity: an outright sale of the asset; a licensing agreement where the economic rights are transferred to the affiliate in exchange for a licensing fee or royalty stream; a sale of services; or a cost sharing agreement, which is an agreement between related entities to share the cost of developing an intangible asset and a proportional share of the rights to the intellectual property that results. A cost sharing agreement typically includes a "buy-in" payment from the affiliate, which supposedly compensates the parent for transferring intangible assets to the affiliate and for incurring the initial costs and risks undertaken in initially developing or acquiring the intangible assets.

The Joint Committee on Taxation has stated that a "principal tax policy concern is that profits may be artificially inflated in low-tax countries and depressed in high-tax countries through aggressive transfer pricing that does not reflect an arms-length result from a related-party transaction."[16] A study by the Congressional Research Service raises the same issue. "In the case of U.S. multinationals, one study suggested that about half the difference between profitability in low-tax and high-tax countries, which could arise from artificial income shifting, was due to transfers of intellectual property (or intangibles) and most of the rest through the allocation of debt."[17] A Treasury Department study conducted in 2007 found the potential for improper income shifting was "most acute with respect to cost sharing arrangements involving intangible assets."[18]

Valuing intangible assets at the time they are transferred is complex, often because of the unique nature of the asset, which is frequently a new invention without comparable prices, making it hard to know what an unrelated third party would pay for a license. According to one recent study by JPMorgan Chase:

> "Many multinationals appear to be centralizing many of their valuable IP [intellectual property] assets in low-tax jurisdictions. The reality is that IP rights are easily transferred from jurisdiction to jurisdiction, and they are often inherently difficult to value."[19]

The inherent difficulty in valuing such assets enables multinationals to artificially increase profits in low tax jurisdictions using aggressive transfer pricing practices. The Economist has described these aggressive transfer pricing tax strategies as a "big stick in the corporate treasurer's tax-avoidance armoury."[20] Certain tax experts, who had previously served in senior government tax positions, have described the valuation problems as insurmountable.[21]

Of various transfer pricing approaches, "licensing and cost-sharing are among the most popular and controversial."[22] The legal ownership is most often not transferred outside the United States, because of the protections offered by the U.S. legal system and the importance of protecting such rights in such a large market; instead, only the economic ownership of certain specified rights to the property is transferred. Generally in a cost sharing agreement, a U.S. parent and one or more of its CFCs contribute funds and resources toward the joint development of a new product.[23] The Joint Committee on Taxation has explained:

> "The arrangement provides that the U.S. company owns legal title to, and all U.S. marketing and production rights in, the developed property, and that the other party (or parties) owns rights to all marketing and production for the rest of the world. Reflecting the split economic ownership of the newly developed asset, no royalties are shared between cost sharing participants when the product is ultimately marketed and sold to customers."[24]

The tax rules governing cost sharing agreements are provided in Treasury Regulations that were issued in December 2011.[25] These regulations were previously issued as temporary and proposed regulations in December 2008. The Treasury Department explained that cost sharing arrangements "have come under intense scrutiny by the IRS as a potential vehicle for improper transfer of taxable income associated with intangible assets."[26] The regulations provide detailed rules for evaluating the compensation received by each participant for its contribution to the agreement[27] and tighten the rules to "ensure that the participant making the contribution of platform intangibles will be entitled to the lion's share of the expected returns from the arrangement, as well as the actual returns from the arrangement to the extent they materially exceed the expected returns."[28] Under these rules, related parties may enter into an arrangement under which the parties share the costs of developing one or more intangibles in proportion to each party's share of reasonably anticipated benefits from the cost shared intellectual asset.[29] The regulations also provided for transitional grandfathering rules for cost sharing entered into prior to the 2008 temporary regulations. As a result of the changes in the regulations, multinational taxpayers have worked to preserve the grandfathered status of their cost sharing arrangements.

C. Transfer Pricing and the Use of Shell Corporations

The Subcommittee's investigations, as well as government and academic studies, have shown that U.S. multinationals use transfer pricing to move the economic rights of intangible assets to CFCs in tax havens or low tax jurisdictions, while they attribute expenses to their U.S. operations, lowering their taxable income at home.[30] Their ability to artificially shift income to a tax haven provides multinationals with an unfair advantage over U.S. domestic corporations; it amounts to a subsidy for those multinationals. The recipient CFC in many cases is a shell entity that is created for the purpose of holding the rights. Shell companies are legal entities without any substantive existence - they have no employees, no physical presence, and produce no goods or services. Such shell companies are "ubiquitous in U.S international tax planning."[31] Typically, multinationals set up a shell corporation to enable it to artificially shift income to shell subsidiaries in low tax or tax haven jurisdictions.

According to a 2008 GAO study, "eighty-three of the 100 largest publicly traded U.S. corporations in terms of revenue reported having subsidiaries in jurisdictions list as tax havens or financial privacy jurisdictions...."[32] Many of the largest U.S. multinationals use shell corporations to hold the economic rights to intellectual property and the profits generated from those rights in tax haven jurisdictions to avoid U.S. taxation.[33] By doing this, multinational companies are shifting taxable U.S. income on paper to affiliated offshore shells. These strategies are causing the United States to lose billions of tax dollars annually.

Moreover, from a broader prospective, multinationals are able to benefit from the tax rules which assume that different entities of a multinational, including shell corporations, act independently from one another. The reality today is that the entities of a parent multinational typically operate as one global enterprise following a global business plan directed by the U.S. parent. If that reality were recognized, rather than viewing the various affiliated entities as independent companies, they would not be able to benefit from creating fictitious entities in tax havens and shifting income to those entities.

In fact, when Congress enacted Subpart F, discussed in detail below, more than fifty years ago in 1962, an express purpose of that law was to stop the deflection of multinational income to tax havens, an activity which is so prevalent today.

D. Piercing the Veil – Instrumentality of the Parent

It has long been understood that a shell corporation could be at risk of being disregarded for U.S. tax purposes "if one entity so controls the affairs of a subsidiary that it 'is merely an instrumentality of the parent.'"[34] Courts have applied the "piercing the corporate veil" doctrine, a common law concept, when determining whether to disregard the separateness of two related entities for corporate and tax liabilities.[35] It is a fact-specific analysis to determine whether the veil of a shell entity should be pierced for tax purposes. The courts over time have looked at such factors as: the financial support of the subsidiary's operations by the parent; the lack of substantial business contacts with anyone except the parent; and whether the property of the entity is used by each as if jointly owned.[36] Despite the availability of this tool to "sham" a corporation and pierce the corporate veil for tax purposes, the IRS and the courts have been hesitant to take action against shell foreign corporations or attribute the activities or income of a CFC to its U.S. parent.[37]

E. Subpart F to Prevent Tax Haven Abuse

As early as the 1960s, "administration policymakers became concerned that U.S. multinationals were shifting their operations and excess earnings offshore in response to the tax incentive provided by deferral."[38] At that time, circumstances were somewhat similar to the situation in the United States today. "The country faced a large deficit and the Administration was worried that U.S. economic growth was slowing relative to other industrialized countries."[39] To help reduce the deficit, the Kennedy Administration proposed to tax the current foreign earnings of subsidiaries of multinationals and offered tax incentives to encourage investments at home.[40]

In the debates leading up to the passage of Subpart F, President Kennedy stated in an April 1961 tax message:

"The undesirability of continuing deferral is underscored where deferral has served as a shelter for tax escape through the unjustifiable use of tax havens such as Switzerland. Recently more and more enterprises organized abroad by American firms have arranged their corporate structures aided by artificial arrangements between parent and subsidiary regarding intercompany pricing, the transfer of patent licensing rights, the shifting of management fees, and similar practices which maximize the accumulation of profits in the tax haven as to exploit the multiplicity of foreign tax systems and international agreements in order to reduce sharply or eliminate completely their tax liabilities both at home and abroad."[41]

Although the Kennedy Administration initially proposed to end deferral of foreign source income altogether, a compromise was struck instead, which became known as Subpart F.[42] Subpart F was enacted by Congress in 1962, and was designed in substantial part to address the tax avoidance techniques being utilized today by U.S. multinationals in tax havens. In fact, to curb tax haven abuses, Congress enacted anti-tax haven provisions, despite extensive opposition by the business community.[43]

F. Subpart F to Tax Current Income

Subpart F explicitly restricts the types of income whose taxation may be deferred, and it is often referred to as an "anti-deferral" regime. The Subpart F rules are codified in tax code Sections 951 to 965, which apply to certain income of CFCs.[44] When a CFC earns Subpart F income, the U.S. parent as shareholder is treated as having received the current income. Subpart F was enacted to deter U.S. taxpayers from using CFCs located in tax havens to accumulate earnings that could have been accumulated in the United States.[45] "[S]ubpart F generally targets passive income and income that is split off from the activities that produced the value in the goods or services generating the income," according to the Treasury Department's Office of Tax Policy.[46] In contrast, income that is generated by active, foreign business operations of a CFC continues to warrant deferral. But, again, deferral is not permitted for passive, inherently mobile income such as royalty, interest, or dividend income, as well as income resulting from certain other activities identified in Subpart F.[47] Income reportable under Subpart F is currently subject to U.S. tax, regardless of whether the earnings have been repatriated. However, regulations, temporary statutory changes, and certain statutory exceptions have nearly completely undercut the intended application of Subpart F.

G. Check-the-Box Regulations and Look Through Rule

"Check-the-box" tax regulations issued by the Treasury Department in 1997, and the CFC "look-through rule" first enacted by Congress as a temporary measure in 2006, have significantly reduced the effectiveness of the anti-deferral rules of Subpart F and have further facilitated the increase in offshore profit shifting, which has gained significant momentum over the last 15 years. Treasury issued the check-the-box regulations which became effective on January 1, 1997. Treasury stated at the time that the regulations were designed to simplify tax rules for determining whether an entity is a corporation, a partnership, a sole proprietorship, branch or disregarded entity (DRE) for federal tax purposes.[48] The regulations eliminated a multi-factor test in determining the proper classification of an entity in favor of a simple, elective "check-the-box" regime. Treasury explained that the rules were intended to solve two problems that had developed for the IRS. First, the rise of limited liability companies (LLCs) domestically had placed stress on the multi-factor test, which determined different state and federal tax treatment for them. Second, international entity classification was dependent upon foreign law, making IRS classification difficult and complex. Check-the-box was intended to eliminate the complexity and uncertainty inherent in the test, allowing entities to simply select their tax treatment.

The regulations, however, had significant unintended consequences and opened the door to a host of tax avoidance schemes. Under Subpart F, passive income paid from one separate legal entity to another separate legal entity – even if they were both within the same corporate structure – was immediately taxable. However, with the implementation of the check-the-box regulations, a U.S. multinational could set up a CFC subsidiary in a tax haven and direct it to receive passive income such as interest, dividend, or royalty payments from a lower tiered related CFC without it being classified as Subpart F income. The check-the-box rule permitted this development, because it enabled the multinational to choose to have the lower tiered CFC disregarded or ignored for federal tax purposes. In other words, the lower tiered CFC, although it was legally still a separate entity, would be viewed as part of the higher tiered CFC and not as a separate entity for tax purposes. Therefore, for tax purposes, any passive income paid by the lower tiered entity to the higher tiered CFC subsidiary would not be considered as a payment between two legally separate entities and, thus, would not constitute taxable Subpart F income. The result was that the check-the-box regulations enabled multinationals for tax purposes to ignore the facts reported in their books – which is that they received passive income. Similarly, check-the-box can be used to exclude other forms of Subpart F income, including Foreign Base Company Sales Income, discussed below.

Recognizing this inadvertent problem, the IRS and Treasury issued Notice 98-11on February 9, 1998, reflecting concerns that the check-the-box regulations were facilitating the use of what the agencies refer to as "hybrid branches" to circumvent Subpart F. "The notice defined a hybrid branch as an entity with a single owner that is treated as a separate entity under the relevant tax laws of a foreign country and as a branch (i.e., DRE) of a CFC that is its sole owner for U.S. tax purposes."[49] The Notice stated: "Treasury and the Service have concluded that the use of certain hybrid branch arrangements [described in Examples 1 and 2 of the Notice] is contrary to the policies and rules of subpart F. This notice (98-11) announces that Treasury and the Service will issue regulations to address such arrangements."[50]

On March 26, 1998, Treasury and the IRS proposed regulations to close the loophole opened by the check-the-box rule to prevent the unintended impact to Subpart F. Recognizing that neither had the authority to change the tax law, the IRS and Treasury stated in the proposed rule "the administrative provision [check-the-box] was not intended to change substantive law. Particularly in the international area, the ability to more easily achieve fiscal transparency can lead to inappropriate results under certain provisions [of subpart F] of the Code."[51]

As noted by the Joint Committee on Taxation, "The issuance of Notice 98-11 and the temporary and proposed regulations provoked controversy among taxpayers and members of Congress."[52] On July 6, 1998, Treasury and the IRS reversed course in Notice 98-35, withdrawing Notice 98-11 and the proposed regulations issued on March 26, 1998. The agencies reversed course despite their expressed concern that the check-the-box rules had changed substantive tax law as set out in Subpart F. The result left the check-the-box loophole open, providing U.S. multinationals with the ability to shift income offshore without the threat of incurring Subpart F taxation on passive foreign income.

Because the check-the-box rule was a product of Treasury regulations and could be revoked or revised at any time, proponents of the rule urged Congress to enact supporting legislation. In 2006, Congress eliminated related party passive income generally from subpart F when it enacted Section 954(c)(6) on a temporary basis. This Section was enacted into law without significant debate as part of a larger tax bill.[53] It provided "look-through" treatment for certain payments between related CFCs, and became known as the CFC look-through rule. It granted an exclusion from Subpart F income for certain dividends, interest, rents and royalties received or accrued by one CFC from a related CFC. As one analyst has explained:

> "*Section 954(c)(6)* came into the law somewhat quietly, through an oddly named piece of legislation (the Tax Increase Prevention and Reconciliation Act of 2005, or TIPRA, which was enacted in May 2006). *Section 954(c)(6)* had earlier passed the Senate and the House as part of the American Jobs Creation Act of 2004, but was then dropped without explanation in conference. When it reemerged one-and-a-half years later in TIPRA it did not attract huge pre-enactment attention, and when finally enacted, its retroactive effective date surprised some taxpayers."[54]

The 2006 statutory look-through provision expired on December 31, 2009, but was retroactively reinstated for 2010, and extended through 2011, by the Tax Relief, Unemployment Insurance Reauthorization, and Job Creation Act of 2010, enacted on December 17, 2010. It was then retroactively reinstated again for 2012, and extended through December 31, 2013 by the American Taxpayer Relief Act, enacted on January 2, 2013.

In addition to the regulations and temporary statutory provisions that have undercut Subpart F's effort to tax offshore passive income, certain statutory exceptions have also weakened important provisions of the law. Two of those exceptions relevant to the Subcommittee's review of Apple are the "same country exception" and "manufacturing exception."

H. Foreign Personal Holding Company Income – Same Country Exception

A major type of taxable Subpart F offshore income is referred to in the tax code as Foreign Personal Holding Company Income (FPHC).[55] It consists of passive income such as dividends, royalties, rents and interest.[56] One example of FPHC income that is taxable under Subpart F is a dividend payment made from a lower tiered to a higher tiered CFC. Another example would be a royalty payment made from one CFC to another. Under Subpart F, both types of passive income received by the CFCs are treated as taxable income in the year received for the U.S. parent.

There are several exceptions, however, to current taxation of FPHC income under Subpart F.[57] One significant exclusion exists for certain dividends, interest and royalties where the payor CFC is organized and operating in the same foreign country as the related CFC recipient. This exclusion is often referred to as the "same country exception." The purpose of this exception is to shield from taxation a payment from one related CFC to another in the same country, on the theory that since both CFCs are subject to the same tax regime, they would have little incentive to engage in tax transactions to dodge U.S. taxes.

I. Foreign Base Company Sales Income – Manufacturing Exception

A second key type of taxable Subpart F offshore income is referred to in the tax code as Foreign Base Company Sales (FBCS) income. FBCS income generally involves a CFC which is organized in one jurisdiction, used to buy goods, typically from a manufacturer in another jurisdiction, and then sells the goods to a related CFC for use in a third jurisdiction, while retaining the income resulting from those transactions. It is meant to tax the retained profits of an intermediary CFC which typically sits in a tax haven. More specifically, FBCS income is income attributable to related-party sales of personal property made through a CFC, if the country of the CFC's incorporation is neither the origin nor the destination of the goods and the CFC itself has not "manufactured" the goods.[58] In other words, for the income to be considered foreign base company sales income, the personal property must be both produced outside the CFC's country of organization and distributed or sold for use outside that same country.[59] The purpose of taxing FBCS income under Subpart F was to discourage multinationals from splitting the manufacturing function from the sales function to deflect sales income to a tax haven jurisdiction.

An exclusion known as the "manufacturing exception" was created, however, for certain FBCS income. Under this exception, the income retained by the intermediary CFC would not be taxed if the CFC itself were a manufacturer and added substantive value to the goods. In 2008, the regulations governing the manufacturing exception were liberalized to make it very easy for a company to claim the exception, further undermining Subpart

F. The 2008 regulations provided that "[a] CFC can qualify for the manufacturing exception if it meets one of three tests. The first two [are] physical manufacturing tests: the substantial transformation test and the substantial activity test. The third test [is] the substantial contribution test."[60] Moving from a requirement that the CFC demonstrate that it performed a manufacturing activity to demonstrating that it made a "substantial contribution" to the goods being sold has transformed this exception into another possible loophole to shield offshore income from Subpart F taxation.

These exceptions and loopholes, as well as other tax provisions, often form overlapping layers of protection against offshore income being taxed under Subpart F. In many instances, a multinational corporation may have multiple exceptions or loopholes available to it to dodge U.S. taxes. For example, as noted above, certain types of passive income may be excluded from Subpart F inclusion through the use of the check-the-box regulations, the look-through rule, or the same country exception. Similarly, FBCS income may be excluded through the use of the check-the-box regulations or the manufacturing exception. If one is not available or taken away, other provisions may be relied on to circumvent the original intent of Subpart F. Through the benefits of deferral and various regulatory and statutory exceptions, the tax code has created multiple incentives for multinational corporations to move income offshore to low or no tax jurisdictions and provided multiple methods to avoid current tax on those offshore transfers. The purpose of the Subcommittee's investigation is to examine those tax loopholes and find an effective way of closing them.

III. APPLE CASE STUDY

A. Overview

The Apple case study examines how Apple Inc., a U.S. corporation, has used a variety of offshore structures, arrangements, and transactions to shift billions of dollars in profits away from the United States and into Ireland, where Apple has negotiated a special corporate tax rate of less than 2%. One of Apple's more unusual tactics has been to establish and direct substantial funds to offshore entities that are not declared tax residents of any jurisdiction. In 1980, Apple created Apple Operations International, which acts as its primary offshore holding company but has not declared tax residency in any jurisdiction. Despite reporting net income of $30 billion over the four-year period 2009 to 2012, Apple Operations International paid no corporate income taxes to any national government during that period. Similarly, Apple Sales International, a second Irish affiliate, is the repository for Apple's offshore intellectual property rights and the recipient of substantial income related to Apple worldwide sales, yet claims to be a tax resident nowhere and may be causing that income to go untaxed.

In addition, this case study examines how Apple Inc. transferred the economic rights to its intellectual property through a cost sharing agreement to two offshore affiliates in Ireland. One of those affiliates, Apple Sales International, buys Apple's finished products from a manufacturer in China, re-sells them at a substantial markup to other Apple affiliates, and retains the resulting profits. Over a four-year period, from 2009 to 2012, this arrangement facilitated the shift of about $74 billion in worldwide profits away from the United States to an offshore entity with allegedly no tax residency and which may have paid little or no income taxes to any national government on the vast bulk of those funds. Additionally, the case study shows how Apple makes use of multiple U.S. tax loopholes, including the check-the-box rules, to shield offshore income otherwise taxable under Subpart F. Those loopholes have enabled Apple, over a four year period from 2009 to 2012, to defer paying U.S. taxes on $44 billion of offshore income, or more than $10 billion of offshore income per year. As a result, Apple has continued to build up its offshore cash holdings which now exceed $102 billion.

B. Apple Background

1. General Information

Apple Inc. is headquartered in Cupertino, California. It was formed as a California corporation on January 3, 1977, and has been publicly traded for more than 30 years.

The current Chairman of the Board is Arthur D. Levinson, Ph.D., and the Chief Executive Officer (CEO) is Tim Cook. Apple is a personal computer and technology company specializing in the design and sale of computers, mobile telephones, and other high-technology personal goods. The sales of personal computers, mobile telephones, and related devices accounts for 95% of Apple's business, while the remaining 5% comes from the sale of related software and digital media.

The company has approximately 80,000 employees worldwide, with 52,000 of those in the United States. The U.S. jobs include 10,000 Apple advisors and 26,000 retail employees. In 2012, Apple reported in its public filings with the Securities and Exchange Commission (SEC) net income of $41.7 billion, based upon revenues of $156.5 billion.[61] These figures translate into earnings per share of $44.15.[62]

Apple conducts its business geographically, with operations for North and South America, including the United States, headquartered in California, and operations for the rest of the world, including Europe, the Middle East, India, Africa, Asia, and the Pacific, headquartered in Ireland.[63] Apple develops its products through research and development conducted primarily in the United States; the materials and components for Apple products are sourced globally.[64]

The finished products are typically assembled by a third party manufacturer in China and distributed throughout the world via distribution centers headquartered in the United States and Ireland.[65]

2. Apple History

Apple was founded in 1976 by Steve Jobs, Steve Wozniak, and Ronald Wayne, to design and sell personal computers.[66] In the late 1970s, Apple decided to expand its presence in Europe and, in the summer of 1980, established several Irish affiliates. Apple entered into a cost-sharing agreement with two of them, Apple Operations Europe (AOE) and its subsidiary, Apple Sales International (ASI).[67] Under the terms of the cost-sharing agreement, Apple's Irish affiliates shared Apple's research and development costs, and in exchange, were granted the economic rights to use the resulting intellectual property. At the time in 1980, Apple's Irish affiliate manufactured the products for sale in Europe.

In December 1980, Apple had its initial public offering of stock and began trading on the New York Stock Exchange.[68] During the 1980s and 1990s, Apple expanded its product lines. While the majority of Apple's research and development continued to be conducted in the United States, its products were manufactured in both California and Cork, Ireland.

By the late 1990s, Apple was experiencing severe financial difficulties and, in 1996 and 1997, incurred two consecutive years of billion-dollar losses. In response, Apple significantly restructured its operations, eliminating many of its product lines and streamlining its offshore operations. In addition, Apple began to outsource much of its manufacturing, using third-party manufacturers to produce the components for the products developed in its California facilities.

Apple also outsourced the assembly of nearly all of its finished products to a third party manufacturer in China. Apple subsequently consolidated its financial management in five shared service centers, with the service center for the Europe region located in Cork, Ireland. It also eliminated over 150 bank accounts in foreign affiliates and established a policy of consolidating excess offshore cash in bank accounts held by its Irish affiliates.

According to Apple, it currently has about $145 billion in cash, cash equivalents and marketable securities, of which $102 billion is "offshore."[69] As of 2011, Apple held between 75 and 100% of those offshore cash assets in accounts at U.S. financial institutions.[70]

C. Using Offshore Affiliates to Avoid U.S. Taxes

Apple continues to organize its sales by dividing them between two regions as it has since 1980. Apple Inc. in the United States is responsible for coordinating sales for the Americas, and Apple's Irish affiliate - Apple Sales International (ASI) is responsible for selling Apple products to Europe, the Middle East, Africa, India, Asia and the Pacific.[71] Apple bifurcates its economic intellectual property rights along these same lines. Apple Inc. is the sole owner of the legal rights to Apple intellectual property. Through a cost-sharing arrangement, Apple Inc. owns the economic rights to Apple's intellectual property for goods sold in the Americas, while Apple's Irish affiliates, Apple Sales International (ASI) and its parent, Apple Operations Europe Inc. (AOE), own the economic rights to intellectual property for goods sold in Europe, the Middle East, Africa, India, and Asia ("offshore").[72] According to Apple, this cost sharing-arrangement enables Apple to produce and distribute products around the world.

Apple Inc. conducts its offshore operations through a network of offshore affiliates. The key affiliates at the top of the offshore network are companies that are incorporated in Ireland and located at the same address in Cork, Ireland. Apple's current offshore organizational structure in Ireland is depicted in the following chart:

Apple's Offshore Organizational Structure

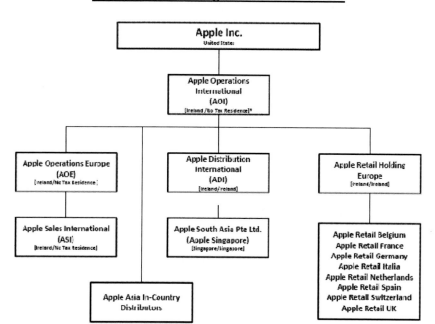

*Listed countries indicate country of incorporation and country of tax residence, respectively.

Prepared by the Permanent Subcommittee on Investigations, May 2013. Source: Materials received from Apple Inc.

1. Benefiting from a Minimal Tax Rate

A number of Apple's key offshore subsidiaries are incorporated in Ireland. A primary reason may be the unusually low corporate income tax rate provided by the Irish government. Apple told the Subcommittee that, for many years, Ireland has provided Apple affiliates with a special tax rate that is substantially below its already relatively low statutory rate of 12 percent. Apple told the Subcommittee that it had obtained this special rate through negotiations with the Irish government.[73] According to Apple, for the last ten years, this special corporate income tax rate has been 2 percent or less:

> "Since the early 1990's, the Government of Ireland has calculated Apple's taxable income in such a way as to produce an effective rate in the low single digits .... The rate has varied from year to year, but since 2003 has been 2% or less."[74]

Other information provided by Apple indicates that the Irish tax rate assessed on Apple affiliates has recently been substantially below 2%. For example, Apple told the Subcommittee that, for the three year period from 2009 to 2011, ASI paid an Irish corporate income tax rate that was consistently below far below 1% and, in 2011, was as low as five-hundreds of one percent (0.05%):

Global Taxes Paid by ASI, 2009-2011

	2011	2010	2009	Total
Pre-Tax Earnings	$ 22 billion	$ 12 billion	$ 4 billion	$ 38 billion
Global Tax	$ 10 million	$ 7 million	$ 4 million	$ 21 million
Tax Rate	0.05%	0.06%	0.1%	0.06%

Source: Apple Consolidating Financial Statements, APL-PSI-000130-232 [Sealed Exhibit].

These figures demonstrate that Ireland has essentially functioned as a tax haven for Apple, providing it with minimal income tax rates approaching zero.

2. Avoiding Taxes by Not Declaring a Tax Residency (a) Apple Operations International (AOI)

Apple's first tier offshore affiliate, as indicated in the earlier chart, is Apple Operations International (AOI). Apple Inc. owns 100% of AOI, either directly or indirectly through other controlled foreign corporations.[75] AOI is a holding company that is the ultimate owner of most of Apple's offshore entities. AOI holds, for example, the shares of key entities at the second tier of the Apple offshore network, including Apple Operations Europe (AOE), Apple Distribution International (ADI), Apple South Asia Pte Ltd. (Apple Singapore), and Apple Retail Europe Holdings, which owns entities that operate Apple's retail stores throughout Europe. In addition to holding their shares, AOI serves a cash consolidation function for the second-tier entities as well as for most of the rest of Apple's offshore affiliates, receiving dividends from and making contributions to those affiliates as needed.[76]

AOI was incorporated in Ireland in 1980.[77] Apple told the Subcommittee that it is unable to locate the historical records regarding the business purpose for AOI's formation, or the purpose for its incorporating in Ireland.[78] While AOI shares the same mailing address as several other Apple affiliates in Cork, Ireland, AOI has no physical presence at that or any other address.[79] Since its inception more than thirty years earlier, AOI has not had any employees.[80] Instead, three individuals serve as AOI's directors and sole officer, while working for other Apple companies. Those individuals currently consist of two Apple Inc. employees, Gene Levoff and Gary Wipfler, who reside in California and serve as directors on numerous other boards of Apple offshore affiliates, and one ADI employee, Cathy Kearney, who resides in Ireland. Mr. Levoff also serves as AOI's sole officer, as indicated in the following chart:[81]

Apple Operations International Officers and Directors

AOI Directors and Officer	Residence	Employer / Job Title
Gene Levoff (Director/Secretary)	USA	Apple Inc./Director of Corporate Law
Gary Wipfler (Director)	USA	Apple Inc./VP and Corporate Treasurer
Cathy Kearney (Director)	Ireland	ADI/VP of European Operations

Source: Apple Response to Subcommittee Questionnaire, APL-PSI-00235

AOI's board meetings have almost always taken place in the United States where the two California board members reside. According to minutes from those board meetings, from May of 2006 through the end of 2012, AOI held 33 board of directors meetings, 32 of which took place in Cupertino, California.[82] AOI's lone Irish-resident director, Ms. Kearney, participated in just 7 of those meetings, 6 by telephone. For a six-year period lasting from September 2006 to August 2012, Ms. Kearney did not participate in any of the 18 AOI board meetings. AOI board meeting notes are taken by Mr. Levoff, who works in California, and sent to the law offices of AOI's outside counsel in Ireland, which prepares the formal minutes.[83]

Apple told the Subcommittee that AOI's assets are managed by employees at an Apple Inc. subsidiary, Braeburn Capital, which is located in Nevada.[84] Apple indicated that the assets themselves are held in bank accounts in New York.[85] Apple also indicated that AOI's general ledger – its primary accounting record – is maintained at Apple's U.S. shared service center in Austin, Texas.[86] Apple indicated that no AOI bank accounts or management personnel are located in Ireland.

Because AOI was set up and continues to operate without any employees, the evidence indicates that its activities are almost entirely controlled by Apple Inc. in the United States. In fact, Apple's tax director, Phillip Bullock, told the Subcommittee that it was his opinion that AOI's functions were managed and controlled in the United States.[87]

In response to questions, Apple told the Subcommittee that over a four-year period, from 2009 to 2012, AOI received $29.9 billion in dividends from lower-tiered offshore Apple affiliates.[88] According to Apple, AOI's net income made up 30% of Apple's total worldwide net profits from 2009-2011,[89] yet Apple also disclosed to the Subcommittee that AOI did not pay any corporate income tax to any national government during that period.[90]

Apple explained that, although AOI has been incorporated in Ireland since 1980, it has not declared a tax residency in Ireland or any other country and so has not paid any corporate income tax to any national government in the past 5 years.[91] Apple has exploited a difference between Irish and U.S. tax residency rules. Ireland uses a management and control test to determine tax residency, while the United States determines tax residency based upon the entity's place of formation. Apple explained that, although AOI is incorporated in Ireland, it is not tax resident in Ireland, because AOI is neither managed nor controlled in Ireland.[92] Apple also maintained that, because AOI was not incorporated in the United States, AOI is not a U.S. tax resident under U.S. tax law either.

When asked whether AOI was instead managed and controlled in the United States, where the majority of its directors, assets, and records are located, Apple responded that it had not determined the answer to that question.[93] Apple noted in a submission to the Subcommittee: "Since its inception, Apple determined that AOI was not a tax resident of Ireland. Apple made this determination based on the application of the central management and control tests under Irish law." Further, Apple informed the Subcommittee that it does not believe that "AOI qualifies as a tax resident of any other country under the applicable local laws."[94]

For more than thirty years, Apple has taken the position that AOI has no tax residency, and AOI has not filed a corporate tax return in the past 5 years. Although the United States generally determines tax residency based upon the place of incorporation, a shell entity incorporated in a foreign tax jurisdiction could be disregarded for U.S. tax purposes if that entity is controlled by its parent to such a degree that the shell entity is nothing more than an instrumentality of its parent. While the IRS and the courts have shown reluctance to apply that test, disregard the corporate form, and attribute the income of one corporation to another, the facts here warrant examination.

AOI is a thirty-year old company that has operated since its inception without a physical presence or its own employees. The evidence shows that AOI is active in just two countries, Ireland and the United States. Since Apple has determined that AOI is not managed or controlled in Ireland, functionally that leaves only the United States as the locus of its management and control. In addition, its management decisions and financial activities appear to be performed almost exclusively by Apple Inc. employees located in the United

States for the benefit of Apple Inc. Under those circumstances, an IRS analysis would be appropriate to determine whether AOI functions as an instrumentality of its parent and whether its income should be attributed to that U.S. parent, Apple Inc.

(b) Apple Sales International (ASI)

AOI is not the only Apple offshore entity that has operated without a tax residency. Apple recently disclosed to the Subcommittee that another key Apple Irish affiliate, Apple Sales International (ASI), is also not a tax resident anywhere. Apple wrote: "Like AOI, ASI is incorporated in Ireland, is not a tax resident in the US, and does not meet the requirements for tax residency in Ireland."[95] ASI is exploiting the same difference between Irish and U.S. tax residency rules as AOI.

ASI is a subsidiary of Apple Operations Europe (AOE) which is, in turn, a subsidiary of AOI.[96] Prior to 2012, like AOI, ASI operated without any employees and carried out its activities through a U.S.-based Board of Directors.[97] Also like AOI, the majority of ASI's directors were Apple Inc. employees residing in California.[98] Of 33 ASI board meetings from May 2006 to March 2012, all 33 took place in Cupertino, California.[99] In 2012, as a result of Apple's restructuring of its Irish subsidiaries, ASI was assigned 250 employees who used to work for its parent, AOE.[100] Despite acquiring those new employees, ASI maintains that its management and control is located outside of Ireland and continues to claim it has no tax residency in either Ireland or the United States.

Despite its position that it is not a tax resident of Ireland, ASI has filed a corporate tax return related to its operating presence in that country.[101] As shown in an earlier chart, ASI has paid minimal taxes on its income. In 2011, for example, ASI paid $10 million in global taxes on $22 billion in income; in 2010, ASI paid $7 million in taxes on $12 billion in income. Those Irish tax payments are so low relative to ASI's income, they raise questions about whether ASI is declaring on its Irish tax returns the full amount of income it has received from other Apple affiliates or whether, due to its non-tax resident status in Ireland, ASI has declared only the income related to its sales to Irish customers. Over the four year period, 2009 to 2012, ASI's income, as explained below, totaled about $74 billion, a portion of which ASI transferred via dividends to its parent, Apple Operations Europe. ASI, which claims to have no tax residence anywhere, has paid little or no taxes to any national government on that income of $74 billion.

3. Helping Apple Inc. Avoid U.S. Taxes via a Cost-Sharing Agreement

In addition to shielding income from taxation by declining to declare a tax residency in any country, Apple Inc.'s Irish affiliates have also helped Apple avoid U.S. taxes in another way, through utilization of a cost-sharing agreement and related transfer pricing practices. Three key offshore affiliates in this effort are ASI, its parent AOE, and Apple Distributions International (ADI), each of which holds a second or third tier position in Apple's offshore structure in Ireland. All three companies are incorporated and located in Ireland, and share the same mailing address. Another key second-tier player is Apple South Asia Pte. Ltd., a company incorporated and located in Singapore (Apple Singapore). These offshore affiliates enable Apple Inc. to keep the lion's share of its worldwide sales revenues out of the United States and instead shift that sales income to Ireland, where Apple enjoys an unusually low tax rate and affiliates allegedly with no tax residency.

The key roles played by ASI and AOE stem from the fact they are parties to a research and development cost-sharing agreement with Apple Inc., which also gives them joint ownership of the economic rights to Apple's intellectual property offshore.[102] As of 2012, AOE had about 400 employees and conducted a small amount of manufacturing in Cork, Ireland involving a line of specialty computers for sale in Europe.[103] Also as of 2012, ASI moved from zero to about 250 employees who manage Apple's other manufacturing activities as well as its product-line sales.[104] As part of its duties, ASI contracted with Apple's third-party manufacturer in China to assemble Apple products and acted as the initial buyer of those finished goods. ASI then re-sold the finished products to ADI for sales in Europe, the Middle East, Africa, and India; and to Apple Singapore for sales in Asia and the Pacific region.[105] When it re-sold the finished products, ASI charged the Apple affiliates a higher price than it paid for the goods and, as a result, became the recipient of substantial income, a portion of which ASI then distributed up the chain in the form of dividends to its parent, AOE. AOE, in turn, sent dividends to AOI.[106]

Cost Sharing Agreement

The cost-sharing agreement is structured as follows.[107] In the agreement, Apple Inc. and ASI agree to share in the development of Apple's products and to divide the resulting intellectual property economic rights. To calculate their respective costs, Apple Inc. first pools the costs of Apple's worldwide research and development efforts. Apple Inc. and ASI then each pay a portion of the pooled costs based upon the portion of product sales that occur in their respective regions. For instance, in 2011, roughly 40 percent of Apple's worldwide sales occurred in the Americas, with the remaining 60 percent occurring offshore.[108] That same year, Apple's worldwide research and development costs totaled $2.4 billion.[109] Apple Inc. and ASI contributed to these shared expenses based on each entity's percentage of worldwide sales. Apple Inc. paid 40 percent or $1.0 billion, while ASI paid the remaining 60 percent or $1.4 billion.[110]

Distribution Structure

For the majority of Apple products, as mentioned earlier, ASI contracted with a third-party manufacturer in China to assemble the finished goods. The persons who actually negotiated and signed those contracts on behalf of ASI were Apple Inc. employees based in the United States, including an Apple Inc. employee serving as an ASI director.[111] The third-party manufacturer manufactured the goods to fill purchase orders placed by ASI.[112] ASI was the initial purchaser of all goods intended to be sold throughout Europe, the Middle East, Africa, India, Asia, and the Pacific region. The chart below illustrates ASI's distribution structure as of 2012.

APPLE'S CURRENT OPERATING STRUCTURE

Source: Apple chart, prepared by Apple at the Subcommittee's request.

Once ASI took initial title of the finished goods, it resold the goods to the appropriate distribution entity, in most cases without taking physical possession of the goods in Ireland.[113] For sales in Europe, for example, ASI purchased the finished products from the third party manufacturer and sold them to ADI. ADI then resold the products to Apple retail subsidiaries located in various countries around Europe, to third-party resellers, or directly to internet customers. For sales in Asia and the Pacific region, ASI sold the finished goods to Apple Singapore, which then re-sold them to Apple retail subsidiaries in Hong Kong, Japan, and Australia, third party resellers, or directly to internet customers.[114]

Although ASI is an Irish incorporated entity and the purchaser of the goods, only a small percentage of Apple's manufactured products ever entered Ireland. Rather, title was transferred between the third party manufacturer and ASI, while the products were being directly shipped to the eventual country of sale. Upon arrival, the products were resold by ASI to the Apple distribution affiliate that took ownership of the goods. The Apple distribution affiliate then sold the goods to either end customers or Apple retail subsidiaries.[115] Apple's distribution process suggests that the location of its affiliates in Ireland was not integral to the sales or distribution functions they performed. Rather, locating the entities in Ireland seemed primarily designed to facilitate the concentration of offshore profits in a low tax jurisdiction.

Shifting Profits Offshore

By structuring its intellectual property rights and distribution operations in the manner it did, Apple Inc. was able to avoid having worldwide Apple sales revenue related to its intellectual property attributed to itself in the United States where it would be subject to taxation in the year received. Instead, Apple Inc. arranged for a large portion of its worldwide sales revenue to be attributed to ASI in Ireland. As explained earlier, according to Apple, Ireland has provided Apple affiliates with an income tax rate of less than 2% and as low as 0.05%. In addition, given ASI's status as a non-tax resident of Ireland, it may be that ASI paid no income tax at all to any national government on the tens of billions of dollars of Apple sales income that ASI received from Apple affiliates outside of Ireland. If that is the case, Apple has been shifting its profits to its Irish subsidiary that has a tax residence nowhere, not to benefit from Ireland's minimal tax rate, but to take advantage of the disparity between Irish and U.S. tax residency rules and thereby avoid paying income taxes to any national government.

The cost-sharing agreement that Apple has signed with ASI and AOE is a key component of Apple's ability to lower its U.S. taxes. Several aspects of the cost-share agreement and Apple's research and development (R&D) and sales practices suggest that the agreement functions primarily as a conduit to shift profits offshore to avoid U.S. taxes. First, the bulk of Apple's R&D efforts, the source of the intangible value of its products, is conducted in the United States, yet under the cost sharing agreement a disproportionate amount of the resulting profits remain outside of the United States. Second, the transfer of intellectual property rights to Ireland via the cost-sharing agreement appears to play no role in the way Apple conducts its commercial operations. Finally, the cost-sharing agreement does not in reality shift any risks or benefits away from Apple, the multinational corporation; it only shifts the location of the tax liability for Apple's profits.

Almost all of Apple's research activity is conducted by Apple Inc. employees in California. The vast majority of Apple's engineers, product design specialists, and technical experts are physically located in California.[116] ASI and AOE employees conduct less than 1% of Apple's R&D and build only a small number of specialty computers.[117] In 2011, 95% of Apple's research and development was conducted in the United States, [118] making Apple's arrangement with ASI closer to a cost reimbursement than a co-development relationship, where both parties contribute to the intrinsic value of the intellectual property being developed.

However, despite the fact that ASI conducts only de minimis research and development activity, the cost sharing agreement gives ASI the rights to the "entrepreneurial investment" profits that result from owning the intellectual property.[119] According to Apple, over the four year period, 2009 to 2012, ASI made cost-sharing payments to Apple Inc. of approximately $5 billion.[120] ASI's resulting income over those same 3 years was $74 billion, a ratio of more than 15 to one, when comparing its income to its costs.[121] In short, ASI profited in amounts far in excess of its R&D contributions.

Cost Sharing Payments and Pre-Tax Earnings of Apple Sales International (Ireland)

	Cost Sharing Payments by ASI	Pre-Tax Earnings of ASI
2009	$ 600 million	$ 4 billion
2010	$ 900 million	$ 12 billion
2011	$ 1.4 billion	$ 22 billion
2012	$ 2.0 billion	$ 36 billion
TOTAL	**$ 4.9 billion**	**$ 74 billion**

Cost Sharing Payments and Pre-Tax Earnings of Apple Inc.
(United States)

	Cost Sharing Payments by Apple Inc.	Pre-Tax Earnings of Apple Inc.
2009	$ 700 million	$ 3.4 billion
2010	$ 900 million	$ 5.3 billion
2011	$ 1.0 billion	$ 11 billion
2012	$ 1.4 billion	$ 19 billion
TOTAL	**$ 4.0 billion**	**$ 38.7 billion**

Source: Information supplied to the Subcommittee by Apple, APL-PSI-000129, 000381-384.

In comparison, over the same four years, Apple Inc. paid $4 billion under the cost-sharing agreement and reported profits of $29 billion.[122] Its cost to profits ratio was closer to 7 to one, substantially less advantageous than that of ASI. The figures disclose that Apple's Irish subsidiary, ASI, profited more than twice as much as Apple Inc. itself from the intellectual property that was largely developed in the United States by Apple Inc. personnel. That relative imbalance suggests that the cost-sharing arrangement for Apple Inc. makes little economic sense without the tax effects of directing $74 billion in worldwide sales revenue away from the United States to Ireland, where it undergoes minimal – or perhaps – no taxation due to ASI's alleged non-tax resident status.

Second, Apple's transfer of the economic rights to its intellectual property to Ireland has no apparent commercial benefit apart from its tax effects. The company operates in numerous countries around the world, but it does not transfer intellectual property rights to each region or country where it conducts business. Instead, the transfer of economic rights is confined to Ireland alone, where the company enjoys an extremely low tax rate. When interviewed, Apple officials could not adequately explain why ASI needed to acquire the economic rights to Apple's intellectual property in order for each to conduct its business. In fact, prior to Apple's reorganization in 2012, ASI had no employees. All business decisions were made by ASI's board of directors, which was composed primarily of Apple Inc. employees and held its meetings in Cupertino, California. Apple's CEO, Tim Cook, told the Subcommittee staff that, during his time as Chief Operating Officer of Apple, he was unable to recall any instance where the ownership of intellectual property rights affected Apple's business operations.[123]

Components used in Apple's finished goods are also produced in multiple countries around the world, without regard to where the economic rights to the underlying intellectual property are located, physically or legally. Many of the component elements of Apple's new products are designed by Apple Inc. in the United States and then manufactured by third parties from different geographic areas, including the United States. The vast majority of Apple's finished products are assembled by a third-party manufacturer in China. The Apple components are sourced globally, and the master servicing agreement governing Apple's relationship with the third-party manufacturer in China that assembles Apple's finished products is negotiated by Apple executives in California. Where this manufacturing work is performed and what entities are selected to perform that work do not appear to be driven by or restricted by which Apple entity holds the economic rights or by where those rights are located.

For example, Apple has noted that the "engine," or central processing unit (CPU), for Apple's iPhones and iPads, is the A5 series of microprocessors built in Austin, Texas. Technically, as a result of Apple's cost-sharing agreement, Apple Inc. owns all of the intellectual property rights (both legal and economic rights) embedded in the CPUs used in the Americas, and ASI owns the intellectual property economic rights for the CPUs used in rest of the world.[124] However, a single facility in Texas produces all of the microprocessors used in all Apple products sold around the world. No business distinction is made between microprocessors manufactured for eventual use in U.S. products, where Apple Inc. owns the intellectual property economic rights, versus use in offshore products, where ASI owns the intellectual property economic rights. In an interview with the Subcommittee, Mr. Cook noted that based on his experience as Chief Operating Officer he considered the costs of Apple components to be borne by the worldwide company rather than the economic rights holders.[125]

Finally, the cost-sharing agreement does not assign any costs, risks, or rewards to any third party independent of Apple. To the contrary, Apple and its offshore affiliates collectively share the risks and rewards of the corporation's research and sales activities. Although Apple Inc. and ASI are distinct legal entities, Apple executives interviewed by the Subcommittee said they viewed the "priorities and interests" of Apple's closely held entities to align with those of Apple Inc.[126] Apple's offshore affiliates operate as one worldwide enterprise, following a coordinated global business plan directed by Apple Inc. In fact, the last two versions of Apple's cost-sharing agreement were signed by Apple Inc. U.S.-based employees, each of whom worked for multiple Apple entities, including Apple Inc., ASI, and AOE.[127] Regardless of where the costs associated with the cost sharing agreement were assigned within the Apple network, or which Apple entities purchased or sold the resulting Apple products, all of the profits and losses from Apple sales were ultimately consolidated in the financial statements of Apple, Inc. The cost sharing agreement did not alter any of those arrangements in any meaningful way. The agreement primarily affects how Apple's R&D costs and sales revenues will be attributed among the affiliates of the international company and in what proportions. Apple, in every case, entered into an agreement with its own entities. In other words, the true function of the cost-sharing agreement has been, not to divide R&D costs with an outside party, but instead to afford Apple the opportunity to direct its costs and profits to affiliates in a low-tax jurisdiction.

These facts raise questions as to whether Apple's intellectual property transfers to related parties perform any function other than to shift profits and tax liability out of the United States to a low-tax jurisdiction.

D. Using U.S. Tax Loopholes to Avoid U.S. Taxes on Offshore Income

Apple's cost-sharing agreement enabled Apple Inc. to direct the lion's share of its worldwide sales income from various Apple affiliates away from the United States to its Irish affiliate, ASI, and its primary offshore holding company, AOI. Because under the U.S. tax code, that offshore income could, under certain circumstances, become subject to U.S. tax in the year received and lose its ability for those taxes to be deferred, Apple took additional steps to shield that income from U.S. taxation.

As noted above, although the United States taxes domestic corporations on their worldwide income, the U.S. tax code allows companies to defer taxes on active business income until that income is returned to the United States. To curb abuse of this foreign income deferral regime, however, Subpart F of the tax code requires that U.S. companies pay tax immediately on certain types of sales revenue transferred between CFCs and on passive foreign income such as dividends, royalties, fees, or interest payments. As explained earlier, the purpose of Subpart F is to prevent U.S. companies from shifting income to tax havens to lower their tax rate without engaging in substantive economic activity. At the same time, the effectiveness of Subpart F has been severely weakened by certain regulations, temporary statutory changes, and statutory exemptions.

According to figures supplied by Apple, over a four year period from 2009 to 2012, as explained further below, Apple used a number of those tax loopholes to avoid Subpart F taxation of offshore income totaling $44 billion.[128] During that time period, Apple generated two types of offshore income that should have been immediately taxed under Subpart F: (1) foreign base company sales (FBCS) income,[129] which involves the sales income Apple directed to Ireland for no reason other than to concentrate profits there, and (2) foreign personal holding company (FPHC) income,[130] which involves passive foreign income such as dividends, royalties, fees, and interest. Apple avoided U.S. taxation for the entire $44 billion through a combination of regulatory and statutory tax loopholes known as the check-the-box and look-through rules.

The following chart depicts both types of income and how Apple structured its offshore operations to avoid U.S. taxes on both.

Apple's Offshore Distribution Structure

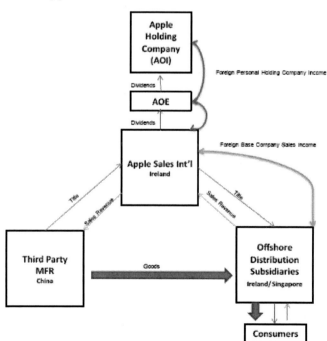

Source: Prepared by Subcommittee based on interviews with Apple employees.

1. Foreign Base Company Sales Income: Avoiding Taxation of Taxable Offshore Income

As explained earlier, foreign base company sales (FBCS) income rules regulate the taxation of goods sold by an entity in one country to a related entity for ultimate use in a different country. The rules were designed to prevent multinational corporations from setting up intermediary entities in tax havens for no purpose except to buy finished goods and sell them to related entities for use in another country in order to concentrate profits from the sales revenue in the tax havens. The distribution structure used by Apple's Irish entities generated significant taxable FBCS income, leading Apple to employ a web of disregarded entities to avoid those U.S. taxes.

The FBCS income designation applies to: (1) purchases of personal property manufactured (by a person other than the CFC) in a jurisdiction other than the country in which the CFC is located, and (2) sold to a related party for use outside of the jurisdiction in which the CFC is located. In the case of Apple, ASI purchased finished Apple goods manufactured in China and immediately resold them to ADI or Apple Singapore which, in turn, sold the goods around the world. ASI did not conduct any of the manufacturing – and added nothing – in Ireland to the finished Apple products it bought, yet booked a substantial profit in Ireland when it resold those products to related parties such as ADI or Apple Singapore.

In fact, ASI never took physical possession of the products it ordered from the third party manufacturer. Transfer was made in title only while the products were being shipped to the country of sale.[131] For example, Apple products sold in Asia were not shipped to Ireland from the third-party manufacturer and then shipped back to Asia for sale. Rather, ASI took title to the manufactured products while they were being shipped to Apple's Asian distribution centers.[132] When they arrived, ASI sold the products to Apple Singapore at a substantial profit.[133] Apple Singapore then resold the products, in turn, to Apple retail entities or end customers.[134] In other instances, the Apple products were shipped directly from the third-party manufacturer to end customers without any Apple intermediary taking prior physical possession.[135]

Transferring title in this manner allowed Apple to retain most of its profits in Ireland, where it has negotiated a favorable tax rate and maintains entities claiming to have no tax residence in any country, and limit the income it reported in the non-tax haven countries where the company did most of its business. For example, in 2011, Apple reported $34 billion in income before taxes; however, just $150 million of those profits, a fraction of one percent, were recorded for Apple's Japanese subsidiaries, even though Japan is one of

Apple's strongest foreign markets.[136] ASI, meanwhile, reported $22 billion in 2011 net income.[137] Those figures indicate that Apple's Japanese profits were being shifted away from the United States to Ireland, where Apple had negotiated a minimal tax rate and maintained two non-tax resident corporations.

It is this type of transfer of worldwide sales income to a tax haven subsidiary that the FBCS income provisions were designed to tax, because they do not contribute to the manufacturing or sales processes, but serve only to concentrate profits in a low tax jurisdiction. Under Subpart F, ASI's income should have been treated as FBCS income subject to U.S. taxation in the year received. Rather than declare that income, however, Apple used the checkthe-box loophole to avoid all U.S. taxation of that FBCS income. When asked to calculate the total amount of U.S. taxes on FBCS income that Apple Inc. was able to avoid by using the check-the-box loophole, Apple provided the following estimates:

Estimated U.S. Taxes Avoided by Apple Inc. Using Check-The Box 2001-2012

	Foreign Base Company Sales Income	Tax Avoided	Tax Avoided Per Day
2011	$ 10 billion	$ 3.5 billion	$ 10 million
2012	$ 25 billion	$ 9.0 billion	$ 25 million
Total	**$ 35 billion**	**$ 12.5 billion**	**$ 17 million**

Source: Information supplied to Subcommittee by Apple, APL-PSI-000386.

These figures indicate that, in two years alone, from 2011 to 2012, Apple Inc. used the checkthe-box loophole to avoid paying $12.5 billion in U.S. taxes or about $17 million per day.

2. Using Check-the-Box to Make Transactions Disappear

To understand how Apple used the check-the-box loophole to avoid those billions of dollars in U.S. tax liability for ASI income, it helps to review Apple's offshore structure as indicated in this chart:

Under the IRS check-the-box regulations, a U.S. multinational can elect to have lower-tier foreign subsidiaries "disregarded" by the IRS as separate legal entities and instead treated as part of an upper-tier subsidiary for tax purposes. If that election is made, transactions involving the disregarded entities disappear for tax purposes, because U.S. tax regulations do not recognize payments made within the confines of a single entity.

Effect of Check the Box

```
┌─────────────────────────────────────────────┐
│                 Apple Inc.                    │
│                 United States                 │
└─────────────────────────────────────────────┘
                      │
        ┌─────────────────────────────┐
        │     Apple Operations          │
        │     International             │
        │        (AOI)                  │
        │  [Ireland /No Tax Residence]* │
        └─────────────────────────────┘
```

┌──────────────────────┐ ┌──────────────────────┐ ┌──────────────────────┐
│ Apple Operations Europe│ │ Apple Distribution   │ │ Apple Retail Holding │
│   (AOE)   │ │ International   │ │ Europe   │
│ [Ireland/No Tax Residence]│ │   (ADI)   │ │ [Ireland/Ireland]   │
│   │ │ [Ireland/Ireland]   │ │   │
└──────────────────────┘ └──────────────────────┘ └──────────────────────┘

┌──────────────────────┐ ┌──────────────────────┐ ┌──────────────────────┐
│ Apple Sales International│ │ Apple South Asia Pte Ltd.│ │ Apple Retail Belgium │
│   (ASI)   │ │ (Apple Singapore)   │ │ Apple Retail France   │
│ [Ireland/No Tax Residence]│ │ [Singapore/Singapore] │ │ Apple Retail Germany │
└──────────────────────┘ └──────────────────────┘ │ Apple Retail Italia   │
                                                      │ Apple Retail Netherlands│
                          ┌──────────────────────┐ │ Apple Retail Spain   │
                          │ Apple Asia In-Country │ │ Apple Retail Switzerland│
                          │ Distributors   │ │ Apple Retail UK   │
                          └──────────────────────┘ └──────────────────────┘

┌──────────────────────┐
│ Disregarded Entities │
└──────────────────────┘

*Listed countries indicate country of incorporation and country of tax residence, respectively.

Prepared by the Permanent Subcommittee on Investigations, May 2013. Source: Materials received from Apple Inc.

In the Apple case, after Apple Inc. makes its check-the-box election, the bottom three tiers of its offshore network – which include AOE, ASI, ADI, Apple Singapore, Apple Retail Holding, and the Apple Retail subsidiaries – all become disregarded subsidiaries of AOI. Those companies are then treated, for U.S. tax purposes, as part of, or merged into, AOI the first tier subsidiary. As a result, the transactions between those disregarded entities are not recognized by the IRS, because the transactions are viewed as if they were conducted within the confines of the same company. The result is that the IRS sees only AOI and treats AOI as having received sales income directly from the end customers who purchased Apple products; that type of active business income is not taxable under Subpart F. The sales income produced when ASI sold Apple products to ADI, Apple Singapore, or Apple's Retail Entities at a substantial markup is no longer considered sales income for tax purposes – it is as if no intercompany sales happened at all. Since no intercompany sales occurred, Subpart F's FBSC income rules no longer applies, which allowed Apple to avoid paying taxes on nearly $44 billion in income from 2009-2012.[138]

3. Using Check-the-Box to Convert Passive Income to Active Income

Apple also uses the check-the-box regulations to avoid U.S. taxation of a second type of offshore income. When an offshore subsidiary of a multinational corporation receives dividends, royalties or other fees from a related subsidiary, that income is considered foreign personal holding company (FPHC) income. That passive income, as it is commonly known, is normally subject to immediate taxation under Section 954(c) of Subpart F. However, once again, under check-the-box rules, if a U.S. multinational elects to have lower-tier subsidiaries "disregarded" – i.e., no longer considered as separate entities – and instead treated as part of an upper-tier subsidiary for tax purposes, any passive income paid by the lower-tier subsidiary to the higher-tier parent would essentially disappear. Because those dividends, royalties and fee payments would be treated as occurring within a single entity, the IRS would not treat them as payments between two legally separate entities or as taxable income under Subpart F.

In Apple's case, in 2011 alone, AOI in Ireland received $6.4 billion in dividends from lower-tier offshore affiliates. Over a four year period, from 2009 to 2012, Apple reported that AOI received a total of $29.9 billion in income, almost exclusively from dividends issued to it by lower-tier CFCs.[139] That dividend income is exactly the type of passive income that Subpart F intended to be immediately taxable. However, by invoking the check-the-box regulations, Apple Inc. was able to designate the lower-tier CFCs as "disregarded entities," requiring the IRS to view them for tax purposes as part of AOI. Once they became part of AOI, their dividend payments became payments internal to AOI and were no longer taxable passive income.

The check-the-box regulations were never intended to be used to convert taxable, offshore, passive income into nontaxable income. Nevertheless, they do, and the resulting loopholes are utilized by Apple and other U.S. multinationals. As explained earlier, the look-through rule provides a similar statutory basis for U.S. multinationals to shield passive offshore income from U.S. taxes. Despite the billions of dollars in offshore income that is escaping U.S. taxation, neither Congress nor the IRS has yet taken any effective action to close these loopholes.

4. Other Tax Loopholes

Even though Apple relies primarily on the check-the-box rules to shield its offshore income from U.S. taxes, if that regulation as well as the look-through rule were eliminated, two other tax loopholes may be available to Apple to continue to avoid Subpart F taxation. They are known as the same country exception and the manufacturing exception.

Same Country Exception

The first loophole is the same country exception.[140] This exception to Subpart F allows payments made between related parties organized and operating within the same country to escape taxation. This exception was created to address the situation in which related entities are located in the same jurisdiction, are theoretically subject to the same tax rate, and supposedly have less incentive to engage in tax-motivated transactions.

Many of the dividends paid to AOI originate from other Apple affiliates incorporated and operating within Ireland, such as AOE and ASI. Under the same country exception, even if the check-the-box and the look-through rules were abolished, the dividend payments made by AOE and ASI to AOI would escape taxation under Subpart F, since the companies are all organized and operating within Ireland. Ironically, because the rule is drafted in terms of the country under whose laws a company is organized, Apple could take advantage of this exception even though it claims AOI, an Irish organized company, is not tax resident in Ireland or anywhere else in the world. Under the explicit terms of the exception, Apple may be able to avail itself of the exception and eliminate all tax liability for intra-country transfers, despite the fact that, according to Apple, AOI and ASI are not tax resident in the same jurisdiction.

Manufacturing Exception

The second loophole is the manufacturing exception to FBCS income.[141] FBCS income is income attributable to related-party sales of personal property made through a CFC if the country of the CFC's incorporation is neither the origin nor the destination of the goods and the CFC itself has not "manufactured" the goods. Under Subpart F, FBCS income is currently taxable. However, under the manufacturing exception, the income from related party purchases and sales will not be characterized as FBCS income if the goods are sold to a related party that transforms or adds substantive value to the goods. In 2008, the regulations governing the manufacturing exception were liberalized to make it very easy for a company to claim such an exception.

Apple told the Subcommittee that it has made no determination about whether the company's supervision of third-party manufacturers qualifies it for the manufacturing exception to FBCS income taxation, since the company relies on the check-the-box rules. However, according to experts consulted by the Subcommittee, the low threshold of the new manufacturing exception rules makes it easy to meet the exception requirements and could be used to avoid taxation.

E. Apple's Effective Tax Rate

When confronted with evidence of actions taken by the company to shield billions of dollars in offshore income from U.S. taxation – including by claiming its offshore Irish subsidiaries, AOI and ASI, have no tax residence in any country and by using the check-the-box and look-through rules to shield its offshore income to from taxation – one of Apple's responses has been to claim that it already pays substantial U.S. tax.[142] Apple's public filings to investors cite an effective tax rate of between 24 and 32 percent. The Subcommittee's investigation has determined, however, that Apple has actually paid billions less to the government than the tax liability reported to investors.

From 2009 to 2012, in its annual report to investors, Apple claimed effective tax rates of between 24% and 32%.[143] In 2011, for example, Apple's annual report (Form 10-K) stated that its net income before taxes was $34.2 billion and that its provision for the payment of corporate income taxes – the company's tax liability – was $8.2 billion, resulting in an effective tax of 24.2%.[144] Apple's calculation, however, included not just its U.S. income taxes, but state and foreign taxes as well. A breakdown of its figures shows that, by its own admission, its effective tax rate for U.S. corporate income taxes was 20.1%, a third lower than the federal statutory rate of 35 percent.

The table below shows Apple's stated provision for income taxes in 2011, broken out by its U.S. federal tax liability, U.S. state-level tax liability, and foreign tax liability[145] as follows:

Apple's Provision for Income Tax in its 2011 Annual Report

	2011 Tax Provision (in millions of dollars)	Effective Tax Rate
Federal tax liability:		
Current	$ 3,884	
Deferred	$ 2,998	
	$ 6,882	**20.1%**
State tax liability:		
Current	$ 762	
Deferred	$ 37	
	$ 799	**2.3%**
Foreign tax liability:		
Current	$ 769	
Deferred	($ 167)	
	$ 602	**1.8%**
Provision for Income Taxes	**$ 8,283**	**24.2%**

Source: Apple Inc. Annual Report (Form 10-K), at 62 (10/26/2011).

Apple calculates its effective tax rate in accordance with GAAP using information in its publicly available annual reports. If the focus, however, were to turn to Apple's federal tax returns and the taxes Apple actually paid to the U.S. treasury each year, its tax payments fall substantially. As part of its investigation, the Subcommittee asked Apple to report the corporate income taxes it actually paid to the U.S. treasury over a three-year period, from 2009 to 2011. According to Apple, the company actually paid just $2.4 billion in federal taxes in 2011, which is $1.4 billion or 30 percent less than the current federal tax provision and $4.4 billion less than the total tax provision included in the company's 2011 annual statement.[146]

While legitimate reasons may exist for differences between a corporation's financial statements and its tax returns, the Subcommittee found large and growing differences in each of the three years it examined with respect to Apple. In all three years, Apple reported much higher provisions for tax on its annual report than it did on its federal tax return for the same year. Moreover, the differences widened substantially over the three-year period, expanding from a 2009 difference of $1.4 billion to a 2011 difference of $4.4 billion. The following chart summarizes that information:

U.S. Tax Liability Reported by Apple Inc. in its Annual Report versus Federal Tax Return, 2009-2011

Form (in millions of dollars)	FY2009	FY2010	FY2011
Total Federal Tax Provision (current plus deferred) reported on 10-K annual report filed with SEC	$ 3.0 billion	$ 3.8 billion	$ 6.9 billion
U.S. tax reported paid on Form 1120 tax return filed with the IRS	$ 1.6 billion	$ 1.2 billion	$ 2.5 billion
Difference:	$ 1.4 billion	$ 2.6 billion	$ 4.4 billion

Source: Information supplied to the Subcommittee by Apple, APL-PSI-000082; Apple Inc. Form 10-K for the fiscal year ended September 29, 2011, at 63.

Tax payments of $1.6 billion, $1.2 billion, or even $2.5 billion produce effective tax rates well below the statutory tax rate. In that, Apple is far from alone. Recent studies indicate that, over a three-year period, from 2008 to 2010, U.S. corporations paid effective tax rates ranging from 12 to 18 percent.[147] One recent study found that 30 large corporations paid no tax at all during a three year period, 2008 to 2010.[148] U.S. records indicate that, in 2011, U.S. corporations collectively paid about $181 billion in federal taxes, compared to the $819 billion in payroll taxes and $1.1 trillion in individual income taxes.[149] Closing offshore tax loopholes such as those created by the check-the-box and look-through rules, the same country exception, and the manufacturing exception, as well as putting a stop to corporations that deny tax residence in any jurisdiction, would help ensure that U.S. multinational corporations begin to pay their share.

The benefits of offshore tax deferral are enhanced by the fact that Apple is able to direct its offshore earnings to jurisdictions with low tax rates. As explained earlier, Apple consolidates as much of its offshore earnings as possible in Ireland, where Apple has an Irish tax rate of less than 2%.[150] Furthermore, Apple's ability to avoid Subpart F taxation through vehicles like check-the-box enables the company to not only shift profits out of the United States, but to shift profits out of other developed countries as well. In 2011, for example, Apple's ability to pass title to the goods it sells around the world through Ireland resulted in 84% of Apple's non-U.S. operating income being booked in ASI.[151] This left very small earnings, and correspondingly small tax liabilities, in countries around the world. In 2011, for example, only $155 million in earnings before taxes were recorded in Apple's UK affiliates. Apple also had no tax liability in its French and German retail affiliates that same year. Through this foreign profit shifting, Apple is able to reduce its foreign tax rate to below 2%.[152] The ability to pay taxes of less than 2% on all of Apple's offshore income gives the company a powerful financial incentive to engage in convoluted tax planning to avoid paying U.S. taxes. Congress can change those incentives by closing offshore tax loopholes and strengthening U.S. tax law.

End Notes

[1] See, e.g., U.S. Senate Permanent Subcommittee on Investigations, "Fishtail, Bacchus, Sundance, and Slapshot: Four Enron Transactions Funded and Facilitated by U.S. Financial Institutions," S.Prt. 107-82 (Jan. 2, 2003); "U.S. Tax Shelter Industry: The Role of Accountants, Lawyers, and Financial Professionals," S.Hrg. 108-473 (No. 18 and 20, 2003); "Tax Haven Abuses: The Enablers, The Tools and Secrecy," S.Hrg 109-797 (Aug. 1, 2006); "Tax Haven Banks and U.S. Tax Compliance," S.Hrg. 110-614 (July 17 and 25, 2008); "Tax Haven Banks and U.S. Tax Compliance: Obtaining the Names of U.S. Clients with Swiss Accounts," S.Hrg. 111-30 (Mar. 4, 2009); "Repatriating Offshore Funds: 2004 Tax Windfall for Select Multinationals," S.Prt. 112-27 (Oct. 11, 2011); and "Offshore Profit Shifting and the U.S. Tax Code – Part 1 (Microsoft and Hewlett-Packard)," S.Hrg.112-*** (Sept. 20, 2012).

[2] 12/8/2011"Reasons for the Decline in the Corporate Tax Revenues" Congressional Research Service, Mark P. Keightley, at.1. See also 4/2011"Tax Havens and Treasure Hunts," Today's Economist, Nancy Folbre.

[3] 4/26/2011 "Parking Earnings Overseas," Zion, Varsheny, Burnap: Credit Suisse, at 3.

[4] 5/1/2013 Audit Analytics, "Foreign Indefinitely Reinvested Earnings: Balances Held by the Russell 3000."

[5] See, e.g., 6/5/2010 "Tax Havens: International Tax Avoidance and Evasion," Congressional Research Service, Jane Gravelle, at 15 (citing multiple studies).

[6] 2/2012 "Foreign Taxes and the Growing Share of U.S. Multinational Company Income Abroad: Profits, Not Sales, are Being Globalized," Office of Tax Analysis Working Paper 103, U.S. Department of Treasury, Harry Grubert, at 1.

[7] 6/5/2010 "Tax Havens: International Tax Avoidance and Evasion," Congressional Research Service, Jane Gravelle, at 14.

[8] 5/16/2012 "Global Tax Rate Makers," JPMorgan Chase, at 2 (based on research of SEC filings of over 1,000 reporting issuers).

[9] 4/23/2013 Apple Second Quarter Earnings Call, Fiscal Year 2013, http://www.nasdaq.com /aspx/calltranscript.aspx?StoryId=1364041&Title=apple-s-ceo-discusses-f2q13-results-earnings-call-transcript.

[10] Subcommittee interview of Apple Chief Executive Officer Tim Cook (4/29/2013).

[11] See 12/2000 "The Deferral of Income Earned through U.S. Controlled Foreign Corporations," Office of Tax Policy, U.S. Department of Treasury, at 12.

[12] 7/20/2010 "Present Law and Background Related to Possible Income Shifting and Transfer Pricing," Joint Committee on Taxation, (JCX-37-10), at 7.

[13] 5/16/2012 "Global Tax Rate Makers," JP Morgan Chase, at 1; see also 4/26/11"Parking Earnings Overseas," Credit Suissc.

[14] See, e.g., U.S. Senate Permanent Subcommittee on Investigations, "Repatriating Offshore Funds: 2004 Tax Windfall for Select Multinationals," S.Rpt. 112-27 (Oct. 11, 2011)(showing that of $538 billion in undistributed accumulated foreign earnings at the end of FY2010 at 20 U.S. multinational corporations, nearly half (46%) of the funds that the corporations had identified as offshore and for which U.S. taxes had been deferred, were actually in the United States at U.S. financial institutions).

[15] See 2/22/2012 "The President's Framework for Business Tax Reform," http://www.treasury. gov/resourcecenter/tax-policy/Documents/The-Presidents-Framework-for-Business-Tax-Reform-02-22-2012.pdf.

[16] 7/20/2010 "Present Law and Background Related to Possible Income Shifting and Transfer Pricing," Joint Committee on Taxation, (JCX-37-10), at 5.

[17] 6/5/2010 "Tax Havens: International Tax Avoidance and Evasion," Congressional Research Service, Jane Gravelle, at 8 (citing 3/2003 "Intangible Income, Intercompany Transactions, Income Shifting and the Choice of Locations," National Tax Journal, vol. 56.2, Harry Grubert, at 221-42).

[18] 7/20/2010 "Present Law and Background Related to Possible Income Shifting and Transfer Pricing," Joint Committee on Taxation, (JCX-37-10), at 7 (citing November 2007 "Report to the Congress on Earnings Stripping, Transfer Pricing and U.S. Income Tax Treaties," U.S. Treasury Department).

[19] 5/16/2012 "Global Tax Rate Makers," JPMorgan Chase, at 1.

[20] 2008 "An Introduction to Transfer Pricing," New School Economic Review, vol. 3.1, Alfredo J. Urquidi, at 28 (citing "Moving Pieces," The Economist, 2/22/2007).

[21] 3/20/2012 "IRS Forms 'SWAT Team' for Tax Dodge Crackdown," Reuters, Patrick Temple-West.

[22] 5/16/2012 "Global Tax Rate Makers," JPMorgan Chase, at 20.

[23] 7/20/2010 "Present Law and Background Related to Possible Income Shifting and Transfer Pricing," Joint Committee on Taxation, (JCX-37-10), at 21.

[24] Id.

[25] Treas. Reg. §1.482-7.

[26] 1/25/2012 "U.S. Department of Treasury issues final cost sharing regulations," International Tax News, Paul Flignor.

[27] 7/20/2010 "Present Law and Background Related to Possible Income Shifting and Transfer Pricing," Joint Committee on Taxation, (JCX-37-10), at 25.

[28] 1/14/2009 "IRS Issues Temporary Cost Sharing Regulations Effective Immediately" International Alert, Miller Chevalier.

[29] 12/12/2012 "Final Section 482 Cost Sharing Regulations: A Renewed Commitment to the Income Method," Bloomberg BNA, Andrew P. Solomon.

[30] U.S. Senate Permanent Subcommittee on Investigations, "Offshore Profit Shifting and the U.S. Tax Code – Part 1 (Microsoft and Hewlett-Packard)," S.Hrg.112-*** (Sept. 20, 2012).

[31] Testimony of Professor Reuven S. Avi-Yonah, hearing before the U.S. Senate Committee on Finance, International Tax Issues, S.Hrg. 112-645 (9/8/2011).

[32] 12/4/2008 "Large U.S. Corporations and Federal Contractors with Subsidiaries in Jurisdictions Listed as Tax Havens or Financial Privacy Jurisdictions," U.S. Government Accountability Office, No. GAO-09-157, at 4.

[33] See, e.g., 2/16/2013 "The price isn't right: Corporate profit-shifting has become big business," The Economist, Special Report.

[34] 2/2011 "Recent IRS determination Highlights Importance of Separation Among Affiliates," by George E. Constantine, at 1, http://www.venable.com/recent-irs-determination-highlights-

importance-of-separation-amongaffiliates-02-24-2011/ (originally published in February 2011 edition of Association Law and Policy, https://www.asaecenter.org/Resources /EnewsletterArticleDetail.cfm?ItemNumber=57654, (citing IRS Priv. Ltr. Rul. 2002-25-046 (Mar. 28, 2002), which cites Moline Properties v. Commissioner of Internal Revenue, 319 U.S. 436, 438 (1943); Britt v. United States, 431 F. 2d 227, 234 (5th Cir. 1970); and Krivo Indus. Supply Co. v. National Distillers and Chem. Corp., 483 F.2d 1098, 1106 (5th Cir. 1973)).

[35] Id. See also, e.g., Moline Properties v. Commissioner of Internal Revenue, 319 U.S. 436, 439 (1943) (holding that, for income tax purposes, a taxpayer cannot ignore the form of the corporation that he creates for a valid business purpose or that subsequently carries on business, unless the corporation is a sham or acts as a mere agent).

[36] Id.

[37] Id. See also Perry Bass v. Commissioner, 50 T.C. 595, 600 (1968) ("[A] taxpayer may adopt any form he desires for the conduct of his business, and ... the chosen form cannot be ignored merely because it results in a tax saving." However, the form the taxpayer chooses for conducting business that results in tax-avoidance "must be a viable business entity, that is, it must have been formed for a substantial business purpose or actually engage in substantive business activity.")

[38] 5/4/2006 "The Evolution of International Tax Policy- What Would Larry Say?" The Laurence Neal Woodworth Memorial Lecture in Federal Tax Law and Policy, Paul Oosterhuis, at 2, http://www.taxanalysts.com/www/features.nsf/articles/3193a0ff95f96d378525726b006f4ad 2?opendocument.

[39] Id.

[40] Id. (citing 1/11/1962 "Annual Message to Congress on the State of the Union," President Kennedy 1 Pub. Papers,

[41] 1961 "President's Recommendations on Tax Revision: Hearings Before the House Ways and Means Committee," reprinted in Richard A. Gordon, Tax Havens and Their Use by United States Taxpayers – An Overview, (2002), at 44.

[42] 5/4/2006 "The Evolution of International Tax Policy- What Would Larry Say?" The Laurence Neal Woodworth Memorial Lecture in Federal Tax Law and Policy, Paul Oosterhuis, at 3, http://www.taxanalysts.com/www/features.nsf/articles/3193a0ff95f96d378525726b006f4ad 2?opendocument.

[43] See, e.g., 12/2000 "The Deferral of Income Earned through U.S. Controlled Foreign Corporations," Office of Tax Policy, U.S. Department of Treasury, at 21.

[44] A CFC is a foreign corporation more than 50% of which, by vote or value, is owned by U.S. persons owning a 10% or greater interest in the corporation by vote ("U.S. shareholders"). "U.S. persons" include U.S. citizens, residents, corporations, partnerships, trusts and estates. IRC Section 957.

[45] See Koehring Company v. United States of America, 583 F.2d 313 (7th Cir. 1978). See also 12/2000 "The Deferral of Income Earned through U.S. Controlled Foreign Corporations," Office of Tax Policy, U.S. Department of Treasury, at xii.

[46] 12/2000 "The Deferral of Income Earned through U.S. Controlled Foreign Corporations," Office of Tax Policy, U.S. Department of Treasury, at xii.

[47] IRC Section 954(c).

[48] IRC Sections 301.7701-1 through 301.7701-3 (1997).

[49] 7/20/2010 "Present Law and Background Related to Possible Income Shifting and Transfer Pricing," Joint Committee on Taxation, (JCX-37-10), at 48.

[50] 1/16/1998, IRS Notice 98-11, at 2.

[51] 3/26/1998 "Guidance Under Subpart F Relating to Partnerships and Branches," 26 CFR Parts 1 and 301 [TD 8767], at 2.

[52] 7/20/2010 "Present Law and Background Related to Possible Income Shifting and Transfer Pricing," Joint Committee on Taxation, (JCX-37-10), at 49.

[53] Tax Increase Prevention and Reconciliation Act of 2005, Pub. L. No. 109-222, § 103(b)(1) (2006).

[54] 4/23/2007 "The New Look-Through [R]ule: W[h]ither Subpart F?" Tax Notes, David Sicular, at 359.

[55] IRC Section 954(c).

[56] IRC Section 954(c).

[57] 7/20/2010 "Present Law and Background Related to Possible Income Shifting and Transfer Pricing," Joint Committee on Taxation, (JCX-37-10), at 36.

[58] IRC Section 954(a)(2). [59]IRC Section 954(d)(1).

[60] 7/20/2010 "Present Law and Background Related to Possible Income Shifting and Transfer Pricing," Joint Committee on Taxation, (JCX-37-10), at 38.

[61] Apple Inc. Annual Report (Form-10K), at 24 (10/21/2012).

[62] Id.

[63] Subcommittee interviews of Cathy Kearney, Apple Distribution International, Vice President of European Operations (4/19/2013) and Tim Cook, Apple Inc.'s former Chief Operating Officer and current Chief Executive Officer (4/29/2013). See also Information supplied to Subcommittee by Apple, APL-PSI-000351.

[64] Subcommittee interviews of Cathy Kearney (4/19/2013) and Tim Cook (4/29/2013).

[65] Id.

[66] See "30 Pivotal Moments In Apple's History," Macworld, Owen W. Linzmayer, (3/30/2006), http://www.macworld.com/article/1050112/30moments.html.

[67] Apple's first cost-sharing agreement was executed on December 1, 1980. See information supplied to Subcommittee by Apple, APL-PSI-000003. AOE was then named Apple Computer Ltd., and ASI was then named Apple Computer International, Inc. Id.

[68] Apple Inc – Frequently Asked Questions, http://investor.apple.com/faq.

[69] 4/23/2013 Apple Second Quarter Earnings Call, Fiscal Year 2013, http://www.nasdaq.com /aspx/calltranscript.aspx?StoryId=1364041&Title=apple-s-ceo-discusses-f2q13-results-earnings-call-transcript.

[70] U.S. Senate Permanent Subcommittee on Investigations, "Offshore Funds Located Onshore," (12/14/2011), at 5 (an addendum to "Repatriating Offshore Funds: 2004 Tax Windfall for Select Multinationals," S.Rpt. 112-27 (Oct. 11, 2011)).

[71] Information supplied to Subcommittee by Apple, APL-PSI-000351.

[72] Id. See also Amended & Restated Cost Sharing Agreement between Apple Inc., Apple Operations Europe, & Apple Sales International, APL-PSI-000020 [Sealed Exhibit].

[73] Subcommittee interview of Phillip Bullock, Apple Inc. Tax Operations Head (5/15/2013).

[74] Information supplied to Subcommittee by Apple, PSI-Apple-02-0004.

[75] Apple Inc. directly owns 97% of AOI and holds the remaining shares through two affiliates, Apple UK which owns 3% of AOI shares, and Baldwin Holdings Unlimited, a nominee shareholder formed in the British Virgin Islands, which holds a fractional share of AOI, on behalf of Apple Inc. Information supplied to Subcommittee by Apple, APL-PSI-000236, and APL-PSI-000352.

[76] Subcommittee interview of Gary Wipfler, Apple Inc. Corporate Treasurer (4/22/2013).

[77] Information supplied to Subcommittee by Apple, APL-PSI-000100. [78]Information supplied to Subcommittee by Apple, APL-PSI-000351.

[79] Subcommittee interview of Cathy Kearney (4/19/2013).

[80] Id.

[81] Mr. Levoff told the Subcommittee that he serves on about 70 different boards of Apple subsidiaries. Subcommittee interview of Gene Levoff, Apple Inc. Director of Corporate Law (5/2/2013). Mr. Levoff also stated that he rarely traveled internationally to carry out his duties as a director on the boards of Apple's subsidiaries, instead carrying out his duties from the United States. Id.

[82] Summary tables of the Board of Directors meetings of AOI prepared by Apple for the Subcommittee, APL-PSI000323, APL-PSI-000341, and APL-PSI-000349.

[83] Subcommittee interview of Gene Levoff (5/2/2013).

[84] Subcommittee interview of Gary Wipfler (4/22/2013).

[85] Id.

[86] Subcommittee interview of Phillip Bullock (11/28/2012).

[87] Id.

[88] Information supplied to Subcommittee by Apple, APL-PSI-000347, APL-PSI-000219, APL-PSI-000181 and APL-PSI-000149.

[89] Apple Consolidating Financial Statements, APL-PSI-000130-232 [Sealed Exhibit].

[90] Information supplied to Subcommittee by Apple, APL-PSI-000240.

[91] Id. Apple reported that, in 2007, AOI paid just under $21,000 in tax in France, related to the sale of a building owned by AOI, and paid a withholding tax on a dividend that same year. Information supplied to Subcommittee by Apple, APL-PSI-000246-247. Apple explained that AOI had a taxable presence in France from 1987-2007, due to its ownership of the building from which it earned rental income until the 2007 sale. Apple has not been able to identify to the Subcommittee any other tax payment by AOI to any national government since 2007.

[92] Information supplied to Subcommittee by Apple, APL-PSI-000241.

[93] "Apple has not made a determination regarding the location of AOI's central management and control. Rather, Apple has determined that AOI is not managed and controlled in Ireland based on the application of the central management and control test under Irish law. The conclusion that AOI is not managed and controlled in Ireland does not require a determination where AOI is managed and controlled." Information supplied to Subcommittee by Apple, APL-PSI-000242.

[94] Information supplied to Subcommittee by Apple, APL-PSI-000239.

[95] Prepared statement of Apple CEO Tim Cook before U.S. Senate Permanent Subcommittee on Investigations, (5/21/2013), at page 14, footnote 8. See also Apple Consolidating Financial Statements, APL-PSI-000130-232 [Sealed Exhibit].

[96] AOI owns 99.99% of AOE and .001% share of ASI; AOE owns 99.99% of ASI. Baldwin Holdings Unlimited, a British Virgin Islands nominee shareholder, holds the remaining fractional share of both AOE and ASI, on behalf of Apple Inc. Information supplied to Subcommittee by Apple, APL-PSI-000236, and APL-PSI-000352.

[97] Subcommittee interview of Tim Cook (4/29/2013); information supplied to the Subcommittee by Apple, APLPSI-000104.

[98] Information supplied to the Subcommittee by Apple, APL-PSI-000343.

[99] Id.

[100] Subcommittee interview of Cathy Kearney (4/19/2013).

[101] See information supplied to Subcommittee by Apple, 5/19/2013 electronic communication ("From 2009 to present, ASI has not met the tax residency requirements in Ireland. However, ASI is an operating company that files an Irish corporate tax return and pays Irish

corporate income tax as required by Ireland. As we indicated in our response to Question 8(c) of our July 6, 2012 submission, ASI's location for tax purposes is Ireland because ASI files a corporate tax return in Ireland. ")

[102] Although AOE and ASI jointly participate in the cost-sharing agreement with Apple Inc., the bulk of Apple's offshore earnings flow to ASI. Information supplied to Subcommittee by Apple, APL-PSI-000384. For simplicity, the Subcommittee will refer to the cost-sharing agreement as between Apple Inc. and ASI, even though the true contractual relationship is between Apple Inc. and both ASI and AOE jointly.

[103] Prior to Apple's restructuring of its Irish affiliates in 2012, all of Apple's 2,452 Irish employees were employed by Apple Operations Europe. In 2012, Apple re-distributed those employees across 5 different Irish affiliates, with the majority now employed by ADI. Information supplied to Subcommittee by Apple, APL-PSI-000103 and PSIApple-02-0002.

[104] Subcommittee interview of Cathy Kearney (4/19/2013).

[105] This description reflects Apple's current distribution arrangements, following its 2012 restructuring of its Irish operations. Prior to the restructuring, ASI contracted with the third party manufacturer, bought the finished Apple products, and then sold those finished products to several Apple retail affiliates and directly to third-party retailers and internet customers. In 2012, Apple split the manufacturing and sales functions so that ASI now arranges for the manufacturing of Apple goods, sells the goods to ADI or Apple Singapore, and ADI or Apple Singapore then manage all sales. As part of this restructuring, Apple moved employees from AOE to ASI and ADI. Information supplied to Subcommittee by Apple, APL-PSI-000103 and PSI-Apple-02-0002??

[106] See, e.g., 11/17/2010 Minutes of a Meeting of the Board of Directors of Apple Operations Europe, APL-PSI000288.

[107] See, e.g., the most recent version of the cost-sharing agreement, 6/25/2009 Amended and Restated Agreement to Share Costs and Risks of Intangibles Development (Grandfathered Cost Sharing Arrangement), APL-PSI-000035 [Sealed Exhibit].

[108] Subcommittee interview of Phillip Bullock (11/28/2012).

[109] Information supplied to Subcommittee by Apple, APL-PSI-000129.

[110] Information supplied to Subcommittee by Apple, APL-PSI-000129.

[111] Information supplied to Subcommittee by Apple, APL-PSI-000392.

[112] Subcommittee interview of Phillip Bullock (11/28/2012).

[113] Prior to 2012, ASI also sold Apple goods directly to end customers or Apple retail entities. Subcommittee interview of Phillip Bullock (11/28/2012).

[114] For sales to China, the third party contract manufacturer sells the finished products to ADI, which then sells to retailers in China. To facilitate this distribution arrangement, ADI sublicenses the rights to distribute Apple products in China for a substantial sum. In FY 2012, for example, ADI paid ASI $5.9 billion for the right to distribute in China. Information supplied to the Subcommittee by Apple, APL-PSI-000234.

[115] Subcommittee interview of Phillip Bullock (11/28/2012). Prior to 2012, ASI sold to Apple retail subsidiaries and directly to internet customers. Since the company reorganized, ASI now sells to ADI and Apple Singapore, and those entities sell to Apple retail subsidiaries, third party resellers, or internet customers. Several Asian subsidiaries also have their own distribution entities that buy from Apple Singapore and resell in country. Id.

[116] Subcommittee interview of Phillip Bullock (5/15/2013).

[117] Information supplied to Subcommittee by Apple, APL-PSI-000233.

[118] Id.

[119] Subcommittee interview of Phillip Bullock (11/28/2012).

[120] Information supplied to Subcommittee by Apple, APL-PSI-000129 and 000382.

[121] Information supplied to Subcommittee by Apple APL-PSI-000384. It is important to note that the cost sharing payments made by ASI have been ongoing for nearly 30 years, and that the costs and resulting profits have fluctuated over that time.

[122] Information supplied to Subcommittee by Apple, APL-PSI-000129 and APL-PSI-000382.

[123] Subcommittee interview of Tim Cook (4/29/2013).

[124] Apple retains the legal rights for the rest of the world. See 6/25/2009 Amended & Restated Agreement to Share Costs and Risks of Intangibles Development (Grandfathered Cost Sharing Arrangement), APL-PSI-000020 [Sealed Exhibit].

[125] Subcommittee interview of Tim Cook, (4/29/2013).

[126] Subcommittee interview of Peter Oppenheimer, Apple Inc. Chief Financial Officer (5/10/2013); Subcommittee interview of Gene Levoff (5/2/2013).

[127] In 2008, Apple Inc, Apple Sales International (ASI), and Apple Operations Europe (AOE) signed an "Amended and Restated Cost Sharing Agreement." The signatory on behalf of AOE, an Irish company, was Gary Wipfler. At the time he was a Board member of both AOE and ASI and was the Treasurer of Apple Inc., in California. The signatory for Apple Inc was Peter Oppenheimer. At the time, he was a board member ASI and AOE, as well as the Chief Financial Officer of Apple Inc. The signatory for ASI, an Irish company, was Tim Cook. At the time, he was a board member of ASI and AOE and the Chief Operating Officer of Apple Inc., in California. In other words, all three signatories to the agreement were directors or officers of all three parties involved in the contract. See Amended & Restated Cost Sharing Agreement Between Apple Inc., Apple Operations Europe & Apple Sales International, May 2008, at15.

In 2009, Apple Inc, ASI and AOE entered into another Cost Sharing agreement which replaced the one signed in 2008. Mr. Oppenheimer, the CFO of Apple Inc. and a director of both ASI and AOE, was the signatory on behalf of Apple Inc. Two other Apple Inc employees signed as directors of ASI and AOE. See Amended and Restated Agreement To Share costs and Risks of Intangibles Development (Grandfathered Cost Sharing Arrangement), June 2009, at19.

[128] Information supplied to Subcommittee by Apple, APL-PSI-000386.

[129] IRC Section 954(d).

[130] IRC Section 954(c).

[131] Subcommittee interview of Cathy Kearney (4/19/13).

[132] Subcommittee interview of Phillip Bullock (11/28/12).

[133] The goods were not necessarily shipped to Singapore either, but may have been shipped to a wide variety of Apple retail entities or end customers across Asia and the Pacific region. Subcommittee interview of Cathy Kearney (4/19/13).

[134] This example is accurate under Apple's current organizational structure. However, Apple Singapore only became an active participant in Apple's distribution channel after Apple's 2012 reorganization. Prior to that reorganization, the same basic structure applied to Apple's distribution channels. At that time, ASI purchased products from the third-party manufacturer and then sold them to Apple affiliates that owned Apple retail stores around the globe. For example, ASI purchase the finished goods from the manufacturer in China and then resold them to an Apple retail store in Australia, with ASI taking ownership of the products while in transit to Australia, then reselling them at a substantial profit to the Apple retail entity upon arrival.

[135] Subcommittee interview of Cathy Kearney (4/19/13).

[136] Apple Consolidating Financial Statements, APL-PSI-000130-232 [Sealed Exhibit].

[137] Id. at APL-PSI-000219.

[138] Information supplied to Subcommittee by Apple, APL-PSI-000386.

[139] Information supplied to Subcommittee by Apple, APL-PSI-000347, APL-PSI-000219, APL-PSI-000181 and APL-PSI-000149.

[140] IRC Section 954(d)(1)(A); Reg. §1.954-3(a)(2).

[141] IRC Section 954(d)(1)(A).

[142] See, e.g., Anna Palmer, *Apple Target of Senate Hearing on Offshore Taxes*, Politico, May, 15, 2013, http://www.politico.com/story/2013/05/apple-hearing-offshore-tax-91425.html.

[143] Apple Inc. Annual Report (Form-10K), at 61 (10/21/2012).

[144] Apple Inc. Annual Report (Form 10-K), at 62 (10/26/2011).

[145] Apple reported an overall tax rate of 24.2%, which is larger than its three component tax rates of 20%, 2.3%, and 1.8%. The larger total is due to U.S. Generally Accepted Accounting Principles (GAAP) which require Apple to include in its "Provision for Income Taxes" all funds it has set aside to pay future taxes, even though Apple continues to retain those funds and has not actually paid those amounts to any tax authority.

[146] Information supplied to Subcommittee by Apple, APL-PSI-000082, referencing data taken from Apple's Form 1120 U.S. Corporation Income Tax Return. According to Apple's 2011 10-K, the company had net excess tax benefits from stock based compensation which is the main reason for the difference between Apple's current tax liability on its financial statement and the liability reported on Apple's tax return. See Apple Inc. Form 10-K for the fiscal year ended September 29, 2011, at 63; Subcommittee interview of Phillip Bullock (5/15/2013).

[147] See, e.g., 2/2012 study by Congressional Budget Office (finding total corporate federal taxes paid fell to 12.1% of profits earned from activities within the United States in FY2011); "Corporate Taxpayers and Corporate Tax Dodgers, 2008-2010," Citizens for Tax Justice and the Institute on Taxation and Economic Policy (11/3/2011), http://www.ctj.org /corporatetaxdodgers/CorporateTaxDodgersReport.pdf.

[148] "Corporate Taxpayers and Corporate Tax Dodgers, 2008-2010," Citizens for Tax Justice and the Institute on Taxation and Economic Policy (11/3/2011), http://www.ctj.org/corporatetaxdodgers/CorporateTaxDodgersReport.pdf.

[149] OMB, Historical Tables, Budget of the U.S. Government. FY2001 (April 2012).

[150] Information supplied to Subcommittee by Apple, PSI-Apple-02-0004.

[151] ASI's operating income was $18 billion in 2011. Apple Consolidating Financial Statements, APL-PSI-000219 [Sealed Exhibit].

[152] According to Apple, in FY2011, its foreign tax rate was 1.8%. See Apple Inc. Annual Report (Form 10-K), at 62 (Oct. 26, 2011).

In: Offshore Profit Shifting ...   ISBN: 978-1-62808-479-5
Editor: Reny Toupin   © 2013 Nova Science Publishers, Inc.

Chapter 2

TESTIMONY OF APPLE INC. HEARING ON "OFFSHORE PROFIT SHIFTING AND THE U.S. TAX CODE – PART 2 (APPLE INC.)"*

I. INTRODUCTION

Apple Inc. ("Apple" or the "Company") appreciates the opportunity to testify before the Permanent Subcommittee on Investigations ("Subcommittee") in connection with its inquiry into the tax practices of multinational companies.

Apple, a California company, employs tens of thousands of Americans, creates revolutionary products that improve the lives of tens of millions of Americans, and pays billions of dollars annually to the US Treasury in corporate income and payroll taxes. Apple's shareholders – from individuals and institutions to pension funds and public employee retirement systems – have benefitted from the Company's success through the appreciation of its stock price and generous dividends. Apple safeguards the capital entrusted to it by its shareholders with prudent management that reflects the Company's extensive international operations. Apple complies fully with both the laws and spirit of the laws. And Apple pays all its required taxes, both in this country and abroad.

---

* This is an edited, reformatted and augmented version of a testimony, presented May 21, 2013 before the Senate Permanent Subcommittee on Investigations.

Apple welcomes an objective examination of the US corporate tax system, which has not kept pace with the advent of the digital age and the rapidly changing global economy. The Company supports comprehensive tax reform as a necessary step to promote growth and enable American multinational companies to remain competitive with their foreign counterparts in both domestic and international markets.

The information Apple has provided to the Subcommittee demonstrates several key points about the Company's operations that are critical to any objective evaluation of its tax practices:

- *Apple has been a powerful engine of job creation in the US.* Apple estimates it has created or supported approximately 600,000 jobs in the US, including nearly 50,000 jobs for Apple employees and approximately 550,000 jobs at other companies in fields such as engineering, manufacturing, logistics and software development. Approximately 290,000 of these American jobs are related to the new "App Economy" launched by Apple's App Store. In less than five years, Apple has paid third-party app developers worldwide over $9 billion in connection with sales of their software to Apple customers.

- *Apple pays an extraordinary amount in US taxes.* Apple is likely the largest corporate income tax payer in the US, having paid nearly $6 billion in taxes to the US Treasury in FY2012. These payments account for $1 in every $40 in corporate income tax the US Treasury collected last year. The Company's FY2012 total US federal cash effective tax rate was approximately 30.5%.[1] The Company expects to pay over $7 billion in taxes to the US Treasury in its current fiscal year. In accordance with US law, Apple pays US corporate income taxes on the profits earned from its sales in the US and on the investment income of its Controlled Foreign Corporations ("CFCs"), including the investment earnings of its Irish subsidiary, Apple Operations International ("AOI").

- *Apple does not use tax gimmicks.* Apple does not move its intellectual property into offshore tax havens and use it to sell products back into the US in order to avoid US tax; it does not use revolving loans from foreign subsidiaries to fund its domestic operations; it does not hold money on a Caribbean island; and it does not have a bank account in the Cayman Islands. Apple has substantial foreign cash because it sells the majority of its products outside the US. International operations accounted for 61% of Apple's revenue last year and two-thirds of its revenue last quarter. These foreign earnings are taxed in the jurisdiction where they are earned ("foreign, post-tax income").

- *Apple carefully manages its foreign cash holdings to support its overseas operations in the best interests of its shareholders.* Apple uses its foreign cash for business operations, geographic expansion, acquisitions and capital investments, and to fund other expenses required by its overseas operations, such as the capital-intensive construction of retail stores in Europe and Asia and the purchase of customized tooling equipment. If the Company repatriated these funds, they would be reduced by a 35% US corporate tax rate. Apple serves its shareholders by keeping these funds overseas where they can be deployed efficiently to fund international operations at a lower cost. As Apple's recent bond issuance demonstrates, the Company can return capital to shareholders using debt at a far lower cost than through repatriation of foreign cash.

- *The dividends distributed among Apple's international affiliates, including AOI, are not subject to US corporate income tax.* AOI and other Apple subsidiaries in Ireland play an important role in the Company's international business activities. Established more than thirty years ago, Apple's base of operations in Ireland now employs nearly 4,000 people engaged in manufacturing, customer service, sales support, supply chain and risk management operations and finance support services. For cash management purposes, these subsidiaries distribute foreign, post-tax income as dividends within Apple's corporate structure. Under US tax law, these foreign intercompany payments are not taxable.

- *Apple's cost sharing agreement with two of its subsidiaries supports high-paying, tax-revenue generating jobs in the US.* Unlike companies that do a substantial share of their research and development in lower cost, foreign jurisdictions, Apple conducts virtually all its R&D in the US. Apple has an agreement with two of its Irish subsidiaries to share the costs and risks of this R&D. The agreement, first established in 1980, is authorized by US law and complies with all US tax regulations. Under the current agreement, the Irish subsidiaries have rights to distribute Apple products in territories outside the Americas in exchange for contributing to jointly-financed R&D efforts in the US. Thus, the agreement supports the funding of the Company's high-paying R&D jobs in the US,

promoting domestic job growth and generating significant tax revenue
for federal and state governments.

- *AOI performs important business functions that facilitate and enhance
  Apple's success in international markets; it is not a shell company.*
  AOI is a holding company that performs centralized cash and
  investment management of Apple's foreign, post-tax income. AOI
  permits Apple to mitigate legal and financial risk by providing
  consolidated, efficient control of its global flow of funds. AOI was
  incorporated in Ireland when Apple began its longstanding business
  presence there, and AOI is properly treated as a CFC under US law.
  The existence of AOI does not reduce Apple's US tax liability.
- *Apple supports comprehensive reform of the US corporate tax system.*
  The Company supports a dramatic simplification of the corporate tax
  system that is revenue neutral, eliminates all tax expenditures, lowers
  tax rates and implements a reasonable tax on foreign earnings that
  allows free movement of capital back to the US. Apple believes such
  comprehensive reform would stimulate economic growth. Apple
  supports this plan even though it would likely result in Apple paying
  more US corporate tax.

II. APPLE'S STORY

Apple is an American success story. Founded by Steve Jobs and Steve
Wozniak nearly four decades ago in a residential garage, Apple has become
the world's most valuable high tech company. Its success results from a simple
priority: Apple strives to make the best products on Earth through a singular
focus on its customers. Apple has introduced new products, new categories –
even new markets – that have profoundly improved people's lives around the
world. True to its California roots, Apple remains headquartered in Cupertino,
and it is now building alarge new campus in that community to accommodate
its substantial growth over the past decade.

Apple designs, manufactures and markets a range of personal computers,
mobile communication and media devices, and portable digital music players.
The Company also provides consumers a variety of related software and
services, including access to third-party digital content and applications. Apple
sells its products worldwide through retail stores, online stores, its direct sales
force, third-party cellular network carriers, wholesalers, retailers and value-
added resellers. The hallmarks for which Apple is best known – creativity,

innovation and design – drive its development activities, almost all of which take place on Apple's main campus in Cupertino.

Apple launched the personal computer revolution in 1976 with the Apple I, followed by the highly successful Apple II. In 1984, Apple reignited that revolution when it introduced its first category-defining product, the Macintosh. With innovations such as the graphical user interface and mouse, the Macintosh made computing accessible to consumers and set the standard for all personal computers that followed.

The mid-1990s proved to be difficult years for the Company. Apple struggled to manage declining sales and market share in an increasingly competitive personal computing market. In 1996 and 1997, Apple lost nearly $2 billion. Many observers predicted Apple would not survive.

Mr. Jobs, who had left Apple in 1985, returned in 1997 with the task of saving the Company. Under his direction, Apple was entirely restructured and focused on innovation. The results are legendary. In 1998, Apple introduced the iMac, a groundbreaking new computer for the consumer market. In 2001, the Company introduced the iPod, another category-defining product that marked Apple's expansion beyond personal computing into the digital marketplace.

Two years later, Apple launched the iTunes on-line music store, changing forever the way consumers legally acquired digital content. The innovative design and customer-focused engineering evident in these products laid the foundation for the Company's explosive growth over the next decade.

In 2007, Apple introduced the iPhone, which quickly set the standard for smartphones. In 2010, Apple introduced the iPad, which established a new market for tablet computers. The iPhone and the iPad illustrate Apple's emphasis on delivering an unmatched user experience and superior technical performance. These products generated unprecedented commercial success and growth for the Company, and created extraordinary value for its shareholders.

In 2008, following the introduction of the iPhone, Apple launched the App Store, which has fundamentally transformed how customers acquire and use software. Today, Apple customers can choose from among 850,000 applications in the App Store. Customers currently download approximately 800 apps per second. Just days ago, the fifty billionth app was downloaded – about seven downloaded apps for every person on Earth.

Apple's growth has created hundreds of thousands of highly-skilled, high-paying jobs for Americans during one of the most difficult economic periods in US history. While the overall size of the domestic workforce has stagnated during the last ten years, Apple has increased its US workforce more than five-fold, from fewer than 10,000 in 2002 to approximately 50,000 today. The Company has also built and opened 250 retail stores in the US. Apple's R&D budget, almost all of which is spent in the US, has also grown dramatically.

Apple is committed to increasing its foundation and operations in the US. The Company is building a new three million square-foot campus in Cupertino that will house 12,000 Apple employees. The Company has broken ground on a new one million square-foot campus in Austin, Texas. In 2010, Apple built one of the country's largest data centers in North Carolina, and it is in the process of constructing two additional data centers in Oregon and Nevada. Reflecting Apple's strong commitment to the environment, these new facilities incorporate green architecture and an emphasis on renewable energy. The North Carolina data center, for example, is powered entirely by renewable energy sources and contains a solar farm and fuel cells on-site, both of which are the largest non-utility owned installations of their kind. The Company will also begin manufacturing one of its Mac lines in the US this year, creating high-quality American manufacturing jobs for a product previously assembled primarily overseas.

Apple's investments over the past decade have resulted in the creation of entirely new products, product categories and industries. The Company estimates that it has created or supported approximately 600,000 jobs for American workers. These US jobs are found in both small and large businesses, and include people who create components for Apple products, deliver those products to Apple's customers and develop apps for sale on the App Store. Apple estimates that approximately 290,000 jobs are related to the "App Economy" created by the App Store.[2]

Apple's commercial success and effective management of cash reserves have yielded significant returns to the Company's shareholders, including individual investors, widely-held mutual funds, US pension funds and public employee retirement systems. Based on the latest available public filings, at least twelve public and private pension funds in the US held Apple stock as their top equity investment, including funds for public employees in Michigan, Ohio and Kentucky.[3] At least twenty-nine such funds identified Apple as a top five holding. All told, these entities own approximately $14.6 billion worth of Apple stock, which entitles them to annual dividend payouts totaling approximately $396 million.[4] At approximately 3% of the S&P 500, Apple is one of the most-widely held equities in the mutual fund industry.

III. APPLE'S CORPORATE STRUCTURE
AND TAX PRACTICES

As a result of its success over the past decade, Apple has likely become the country's largest corporate income taxpayer. In FY2012, Apple made income tax payments to the US Treasury totaling nearly $6 billion – or $16 million per day – and had a US federal cash effective tax rate of approximately 30.5%. Expressed differently, Apple paid $1 out of every $40 of corporate income taxes collected by the US Treasury last year. The Company expects its US income tax bill to increase to more than $7 billion this year.

Income taxes do not represent Apple's entire contribution to the federal and state treasuries. In FY2012, the Company paid approximately $327 million in the employer's share of payroll taxes for its US-based employees and $830 million in income taxes to state governments. Apple also pays a host of other state and local taxes arising from its property holdings and operations in the US. In addition, Apple paid or collected and remitted over $1.3 billion of US state sales and use taxes.

While Apple's success in the US market has continued, the global popularity of its products has soared. The Company's international revenue has outpaced US sales in recent years and substantially contributed to its rapid growth. Last year, approximately 61% of Apple's revenue was derived from its international operations. International revenue accounted for about two-thirds of Apple's revenue last quarter. Revenues from international operations are taxed in accordance with the laws of the countries where they are earned.

As a result of its international success, Apple has accumulated significant amounts of cash outside the US. As described in greater detail below, Apple carefully manages this foreign, post-tax income to support its foreign operations through a corporate structure that protects and promotes the interests of its shareholders. Current US corporate income tax law severely discourages the use of these funds in the US by imposing a 35% tax on repatriation.

To support its global business, Apple relies on a network of foreign subsidiaries incorporated in countries around the world to perform a variety of functions, from manufacturing to sales and support. Several subsidiaries are incorporated in Ireland, where Apple began operations in 1980. The Irish subsidiaries, which are involved in manufacturing, distribution, technical support, sales support and finance support services, include the following: Apple Operations International ("AOI"), Apple Operations ("AO"), Apple

Operations Europe ("AOE"), Apple Sales International ("ASI") and Apple Distribution International ("ADI"). Apple's Irish subsidiaries employ nearly 4,000 people[5] and pay taxes there as required by Ireland. Apple recently broke ground on an expansion of its campus in Cork.

To meet the needs of Apple's expanding overseas operations, the Company's Irish subsidiaries have distributed active foreign, post-tax income as dividend payments within Apple's foreign corporate structure. These dividends represent profit that was previously taxed in accordance with the laws of the local jurisdiction in which it was earned. Under US tax law, these dividends are not taxable. However, in accordance with US Subpart F income rules, Apple Inc. pays taxes to the US Treasury on investment income generated by the assets held by the Irish subsidiaries, including interest earned on their cash.

Apple wants to make clear to the Subcommittee that the Company does not use its Irish subsidiaries or any other entities to engage in the following tax practices that were the focus of the Subcommittee's September 20, 2012 hearing, entitled Offshore Profit Shifting and the US Tax Code. Specifically, Apple does not move its intellectual property into offshore tax havens and use it to sell products back into the US to avoid US tax, nor does it use revolving loans from CFCs to fund its domestic operations. Apple does not hold money on a Caribbean island, does not have a bank account in the Cayman Islands, and does not move any taxable revenue from sales to US customers to other jurisdictions in order to avoid US taxation. Nonetheless, Apple realizes the Subcommittee staff has expressed an interest in its corporate structure and some of its tax-related practices. The Company appreciates the opportunity to address each of the Subcommittee staff's apparent concerns below.

A. Cost Sharing Agreement among Apple Inc., ASI and AOE

Pursuant to US Treasury regulations, Apple Inc. properly uses a cost sharing agreement with two of its Irish subsidiaries to share the R&D costs of co-developing its innovative products for a global market. Cost sharing agreements allow parties to combine financial resources, and therefore jointly bear risk, to invest in R&D in exchange for a share of the rights to any resulting intellectual property for their respective markets. Apple's cost sharing agreement was first put in place in 1980, when Apple had revenue of $117 million and the invention of the iPhone was decades into the future.

Companies commonly use cost sharing agreements for non-tax business purposes. These agreements were sanctioned by the US Congress in 1986 and are expressly authorized by US Treasury regulations.[6] Those rules acknowledge that R&D cost sharing agreements are common between unrelated parties. Accordingly, the regulations explicitly permit related parties, such as wholly-owned subsidiaries, to make use of such arrangements to grant licenses to share the rights to intellectual property that is co-developed under those agreements. By sharing the costs and benefits of R&D activities among domestic and international companies, these agreements allow US multinational companies like Apple to fund high-paying R&D jobs in this country.

Apple's cost sharing agreement is regularly audited by the IRS and complies fully with all applicable Treasury regulations. This agreement allows the Company to co-develop and share the risk of developing new products with its foreign subsidiaries. Under the agreement's terms, ASI and AOE, which are two of Apple's Irish operating companies, partially fund R&D costs incurred by Apple Inc. The share of R&D costs funded by the Irish subsidiaries is based on the relative share of revenue they earn outside the Americas from the intellectual property covered by the agreement. For example, in FY2012, approximately 61% of Apple's revenue was earned internationally, and ASI and AOE funded more than half of Apple's R&D costs. Apple Inc. does not deduct on its US tax return the R&D costs funded by ASI and AOE.

Apple's initial cost sharing agreement was executed in December 1980, when the Company selected Ireland as its principal base of operations for distributing products and servicing customers in western Europe. The cost sharing agreement afforded Apple the means to share the costs and risks of that market expansion with its Irish subsidiaries. In return, the Irish subsidiaries received a license to Apple Inc.'s intellectual property and the right to share in any profits that might result.

When Apple struggled financially and lost market share in the 1990s despite investments in new products and services, the Irish subsidiaries also lost money. The Irish subsidiaries had to fund a portion of Apple's R&D efforts, yet they were not realizing offsetting gains from the sale of Apple products in their markets. The Company almost ran out of cash and was on the verge of bankruptcy.

Eventually, Apple's R&D investments paid off. The R&D funded by Apple Inc., ASI and AOE fueled worldwide commercial success and growth. After paying their share of R&D expenses and bearing losses during some very lean years in the 1990s, Apple's Irish subsidiaries are now profiting from the cost sharing arrangement established three decades ago. This balance of risk and reward is precisely what was contemplated by the US Treasury regulations governing cost sharing agreements.

From a tax policy standpoint, cost sharing agreements play an important role in encouraging companies like Apple to keep R&D efforts – and the high-paying, income tax generating jobs associated with them – in the US. As an American multinational company, Apple is proud of its efforts to create American jobs. Its cost sharing arrangement enables the Company to use revenues earned overseas to fund R&D in the US. Some commentators have urged eliminating these types of cost sharing agreements, but doing so would harm American workers and the broader US economy. If cost sharing agreements were no longer available, many US multinational companies would likely move high-paying American R&D jobs overseas.

B. Apple Operations International

AOI is a holding company that directly or indirectly holds shares in certain Apple foreign operating subsidiaries, including ASI and AOE. A holding company is a widely recognized corporate form under the laws of the US and foreign countries. Some of America's most successful companies, such as Procter & Gamble and Johnson & Johnson, operate as holding companies. AOI functions, as holding companies do, to exercise control over foreign operating subsidiaries on behalf of, and under the direction of, AOI's parent company, Apple Inc. AOI's proper observance of corporate formalities is consistent with this status, as is the appointment of US-based directors who are Apple Inc. employees. These employees act both as AOI directors and stewards for Apple Inc.'s ultimate 100% ownership interest in AOI.

AOI consolidates and manages a substantial portion of Apple's foreign, post-tax income through intercompany dividends. This consolidation creates economies of scale that allow AOI to obtain better rates of return with money management firms. The consolidation of funds into as few bank accounts as possible improves operational controls over cash held within and among other foreign subsidiaries. AOI allows Apple to efficiently redeploy funds to meet the needs of Apple's international operations. Using this structure, Apple's Irish subsidiaries have invested billions of dollars to fund customized tooling equipment used to manufacture Apple products.

The Irish subsidiary structure has also allowed the Company to transfer funds efficiently to construct retail stores in Europe and elsewhere.

AOI uses US-based investment advisors and banks to manage its financial assets. This reflects a prudent business decision regarding the benefits AOI can derive from these service providers. AOI's cash and investments are held in US banks and centrally managed to promote efficiency and offer the opportunity to earn higher returns, which are subject to US income tax. These assets are held in US dollars to mitigate the economic and accounting effects of foreign currency fluctuations. There are severe limitations, however, on Apple's use of these non-repatriated earnings. For example, Apple cannot use these funds to pay US employees, make capital investments in the US, repurchase shares or pay dividends.

AOI invests in US securities for many of the same reasons as other foreign companies: AOI deems these investments most suitable to accomplish its cash management goals of capital preservation and protection against currency fluctuations. US tax law does not interpret these investment-related activities as an indication of deemed repatriation or national corporate residency. Such an interpretation could have a negative impact on US advisors and banks. Foreign companies, for example, might decline to use US-based financial services firms out of concern that such activities would expose them to US taxation. For the same reason, foreign companies might decline to purchase US Government debt, raising the government's borrowing costs.

As Congress affirmed when it codified the economic substance doctrine in the Patient Protection and Affordable Care Act, taxpayers are free to use a domestic or foreign entity for purposes of conducting their foreign affairs.[7] AOI is incorporated in Ireland; thus, under US law, it is not tax resident in the US. AOI is also not tax resident in Ireland because it does not meet the fact-specific residency requirements of Irish law. This does not mean that AOI's income has not been subject to tax. AOI's dividend receipts consist of foreign, post-tax income, i.e., funds that have already been subject to tax in accordance with the laws of the countries where they were earned. AOI's investment income earned on its cash holdings is taxable to Apple Inc., because AOI is a CFC that is wholly owned by Apple Inc.[8]

It should be emphasized that AOI does not reduce Apple's tax bill in the US. If AOI did not exist, the funds it receives from other foreign subsidiaries through dividends would simply remain in the custody of those subsidiaries and would not be subject to US corporate income tax. However, without AOI, Apple would lose the considerable risk management and administrative benefits it provides for the Company's international operations.

C. Deferred Tax Liability

Some observers have suggested that Apple's recording of a US deferred tax liability for portions of its foreign, post-tax income reflects the Company's current plan for cash repatriation. This is incorrect. Apple reports this liability in accordance with a US accounting standard known as APB 23. This recording of a US deferred tax liability provides no indication of the Company's intentions to repatriate foreign, post-tax income. Indeed, Apple has no current plans to repatriate these funds.

IV. APPLE'S CAPITAL RETURN PROGRAM

On April 23, 2013, Apple announced it would substantially increase the return of capital to shareholders. Under this program, Apple expects to return $100 billion to its shareholders in less than three years through a combination of share repurchases and dividends. Apple will expend $60 billion in the share repurchase program, making it the largest single share repurchase authorization in history. Apple's increased quarterly dividend of $3.05 per share makes the Company among the largest dividend payers in the world, with annual payments to shareholders of about $11 billion. Apple expects to fund the capital return program from existing US cash, future cash generated in the US and domestic borrowing.

Some observers have questioned Apple's decision to fund part of its return of capital by issuing $17 billion in debt rather than repatriating some offshore funds. Apple respectfully suggests that any objective analysis will conclude that the Company's choice to issue debt, rather than repatriate foreign earnings, was in its shareholders' best interests. Indeed, the Company's largest investors and financial analysts urged Apple to engage in borrowing to add leverage to its capital structure.

If Apple had used its overseas cash to fund this return of capital, the funds would have been diminished by the very high corporate US tax rate of 35% (less applicable foreign credits). By contrast, given today's historically low interest rates, issuing debt at a cost of less than 2% is much more advantageous for the Company's shareholders. Because Apple was able to borrow at a cost lower than the cost of its equity, issuing debt lowered Apple's overall cost of capital.

Additionally, issuing debt served the interests of Apple's shareholders because the debt's interest rate is lower than the dividend yield on the

Company's equity. Thus, for every debt-financed repurchase of a share of stock, the Company pays less in debt interest than it would have paid in a dividend to the holder of that share. The prudence of this decision has been ratified by the very positive response to Apple's announcement from the investors in its bond offering.

V. APPLE SUPPORTS COMPREHENSIVE CORPORATE TAX REFORM

Apple agrees with those in Congress who believe the current US corporate tax system must be reformed to reflect both the digital age and the globalization of commerce. The Company believes the current system, which applies industrial era concepts to a digital economy, actually undermines US competitiveness.

Apple has always believed in the simple, not the complex. This is evident in the Company's products and the way it conducts itself. In this spirit, Apple has recommended to the Obama Administration and several members of Congress – and suggests to the Subcommittee today – to pass legislation that dramatically simplifies the US corporate tax system. This comprehensive reform should:

- Be revenue neutral;
- Eliminate all corporate tax expenditures;
- Lower corporate income tax rates; and
- Implement a reasonable tax on foreign earnings that allows free movement of capital back to the US.

Apple recognizes these and other improvements in the US corporate tax system may increase the Company's taxes. Apple is not opposed to such a result if it occurs in the context of an overall improvement in efficiency, flexibility and competitiveness. Apple believes the changes it proposes will stimulate the creation of American jobs, increase domestic investment and promote economic growth.

While some Subcommittee members may have differing views on these tax policy matters, Apple hopes the Subcommittee will see that these recommendations aim to create meaningful change and go well beyond what most US companies propose. As both a pioneer and participant in the

American innovation economy, Apple looks forward to working with the Subcommittee on its efforts to encourage comprehensive reform of the US corporate tax system. Apple appreciates the opportunity to appear before the Subcommittee to contribute constructively to this important debate.

End Notes

[1] This calculation is a reflection of federal taxes Apple paid against US pretax earnings, not a calculation of Apple's final tax liability for FY2012.

[2] Apple Inc., Apple - Job Creation, available at http://www.apple.com/about/job-creation/.

[3] These pension funds include the Michigan Department of Treasury, the State Teachers' Retirement System of Ohio and the Kentucky Teachers' Retirement System. The State of Wisconsin Investment Board, the Ohio Public Employees Retirement System and the Arizona State Retirement System each identifies Apple stock as its second largest holding. At a share price of $450 and annual dividend of $12.20 per share, these six funds' combined holdings amount to more than $2.3 billion and entitle them to annual dividend payouts of approximately $62 million.

[4] Assuming a share price of $450 and an annual dividend of $12.20 per share.

[5] The number of employees fluctuates with seasonal demands and new product launches.

[6] H.R. Rep. 99-841, II-638 (1986); Treas. Reg. sec. 1.482-7. Final regulations addressing cost sharing arrangements were first issued in 1968. See 1968 Treas. Reg. sec. 1.482-2(d)(4), 33 Fed. Reg. 5848 (4/16/68). The US Treasury and IRS revised the cost sharing agreement provisions in final regulations in 1995 and 2011. See TD 8632, 60 Fed. Reg. 65553 (12/20/95); T.D. 95568 (12/16/2011, amend T.D. 9569 (12/19/2011).

[7] See General Explanation of Tax Legislation Enacted in the 111th Congress, JCS-2-11, 379 (March 2011).

[8] Like AOI, ASI is incorporated in Ireland, is not tax resident in the US, and does not meet the requirements for tax residency in Ireland. ASI is an operating company with employees who manage the procurement and supply chain for Apple products sold abroad by ADI. Accordingly, ASI files an Irish corporate tax return and pays taxes in Ireland. ASI's investment income is taxed in the US on Apple Inc.'s tax return as Subpart F income. The fact that ASI is not tax resident in Ireland does not reduce Apple Inc.'s US tax liability.

In: Offshore Profit Shifting ...   ISBN: 978-1-62808-479-5
Editor: Reny Toupin   © 2013 Nova Science Publishers, Inc.

Chapter 3

TESTIMONY OF J. RICHARD HARVEY, JR., DISTINGUISHED PROFESSOR, VILLANOVA UNIVERSITY SCHOOL OF LAW. HEARING ON "OFFSHORE PROFIT SHIFTING AND THE U.S. TAX CODE - PART 2 (APPLE INC.)"[*]

Chairman Levin, Ranking Member McCain, and members of the Subcommittee, thank you for the opportunity to testify this morning on issues surrounding the shifting of profits by MNCs to low-tax jurisdictions, and specifically the techniques used by Apple, Inc. This is a very important topic that deserves to be highlighted and discussed.

I am the Distinguished Professor of Practice at the Villanova University School of Law and Graduate Tax Program. Immediately prior to joining the Villanova faculty I was the Senior Advisor to former IRS Commissioner Doug Shulman where my focus was international tax issues and improving corporate tax transparency (e.g., Schedule UTP). I joined the IRS upon retiring from a Big 4 accounting firm as a Managing Partner and my experience also includes prior government service in the US Treasury Department Office of Tax Policy during the negotiation and implementation of the 1986 Tax Reform Act.

With the Chairman's permission, I request that my written testimony be submitted for the record and I will summarize my major observations in oral remarks.

---

[*] This is an edited, reformatted and augmented version of a testimony presented May 21, 2013 before the Senate Permanent Subcommittee on Investigations.

EXECUTIVE SUMMARY

Apple, Inc. (Apple) is an iconic US multinational corporation (MNC) that has enjoyed extraordinary financial success. In addition to demonstrating excellence in designing, building, and selling consumer products, Apple has been very successful at minimizing its global income tax burden. For example:

- Pursuant to a long-standing cost sharing agreement, Apple recorded approximately $22 billion of its 2011 pre-tax income in Ireland.[1] As a result, 64% of Apple's global pre-tax income is recorded in Ireland where only 4% of its employees and 1% of its customers are located.
- If Apple had not entered into the cost sharing agreement, 2011 US pre-tax income would have increased by approximately $22 billion resulting in an additional federal tax liability of approximately $22 billion x 35% = $7.7 billion.
- Despite a published tax rate of 12.5%, Apple negotiated a special tax deal that resulted in only $13 million of Irish tax expense being recorded with respect to the $22 billion of Irish income.
- 60% of Apple's 2011 sales were to customers in countries other than the US and Ireland, but only 6% of the consolidated pre-tax income was recorded in such countries.

Although shifting income out of the US and locating it in a tax haven like Ireland are key steps in Apple's international tax planning, Apple must also avoid the so-called "Subpart F" rules.

These rules were originally designed to tax passive income earned by foreign subsidiaries of US MNCs and therefore discourage the shifting of income out of the US. However, the rules have been substantially "gutted" through adoption of (i) the check-the-box regulations, (ii) the CFC look-through rule, (iii) the contract manufacturing exemption, and to a lesser extent (iv) the same-country exception.

In order to restore the effectiveness of the Subpart F rules and discourage the shifting of income from the US, Congress should quickly adopt the following tax policy recommendations:

- *Substantially restrict the tax planning tools used to circumvent Subpart F* – The check-the-box regulations should be restricted for foreign entities, the CFC look-through rule should at most apply to only dividends, the contract manufacturing exemption should be either eliminated or substantially tightened, and the same-country exception should be modified.

- *Increase transparency* – It is often very difficult for the IRS to get a true picture of a MNC's global tax planning. Thus, US MNCs should be required to report the geographical location of income, tax, and other pertinent information.

Although adoption of these proposals would be very beneficial, additional tax policy changes will also be needed. The reason is that as long as the US (and the rest of the world) applies an arms-length pricing standard to transactions between controlled parties, there will be an opportunity for MNCs to shift income. As a result, a longer-term solution will ultimately be needed. My personal recommendations, in order of preference, are as follows:

- *Obtain global consensus on how to address corporate tax havens* – Unfortunately, this could be very difficult and the US may have to act unilaterally.
- *Substantially lower the corporate tax rate and replace the lost revenue with a VAT or other revenue source* - However, this is likely a political nonstarter.
- *If the arms-length standard is maintained, there needs to be adequate base erosion protections* - For example, a minimum tax should be imposed on earnings recorded in a tax haven, and US deductions for expenses related to foreign income should be restricted (e.g., interest). If a minimum tax is adopted, emphasis should be on making the minimum tax relatively simple to calculate and audit.
- *Consider replacing the arms-length standard with a formula apportionment approach if adequate base erosion protections are not enacted* – However there are major design issues that would need to be addressed.

Selected Apple Information

Per a review of both (i) information supplied by Apple to the Permanent Subcommittee on Investigation (PSI) and (ii) publicly available information in its Form 10-K, a clearer picture emerges about the results of Apple's international tax planning. For example, the table below summarizes the geographic location of 2011 pre-tax income, employees, and customers:

	2011 Pre-Tax Income		Employees @ June 20112		2011 Customer
Country	$ Billions	%	#	$	Location3
United States	10.2	30%	67%	79%	39%
Ireland	22.0	64	4	3	1
Other countries	2.0	6	29	18	60
Consolidated	34.2	100%	100%	100%	100%

In addition, the effective tax rate on all foreign earnings was approximately $600 million/$24 billion = 2.5% with substantially all of the $600 million of tax being incurred on the $2 billion of income earned from foreign countries other than Ireland.

The information below on the profitability of US vs. non-US operations and the allocation of certain expenses between US and non-US operations is also of potential interest:

	Pre-tax Income/Sales	% of Pre-tax Income	General & Administrative Expenses		Sales, Marketing, & Distribution Expenses	
			$ billions	%	$ billions	%
US	24%	30%	1.7	85%	3.3	59%
Non-US	36%	70%	0.3	15%	2.3	41%
Consolidated	32%	100%	2.0	100%	5.6	100%

In addition to the above summary information, there is substantial information summarized in the PSI report prepared for today's hearing that sheds additional light on Apple's international tax planning.

Major Observations Surrounding Apple

Having reviewed the Apple information several major observations can be made,[4] including:

- *Apple received a substantial US tax benefit from a "cost sharing agreement"* - The $22 billion of pre-tax income recorded in Ireland results from a cost sharing agreement whereby Apple transferred to Apple Sales International (ASI), an Irish entity, its development rights to Apple products outside of the Americas.[5] If Apple had not transferred these rights to ASI, 2011 US pretax income would have been approximately $22 billion higher.[6]

- *The overall effective tax rate on Apple's foreign earnings is only 2.5%* - Apple was able to achieve this very low rate through the following:

  o Negotiated Irish tax - In addition to having a disproportionate amount of pre-tax income recorded in Ireland,[7] the effective tax rate charged on the $22 billion of Irish income appears to be less than 1%.[8] Given the stated corporate tax rate in Ireland is 12.5%, it seems very clear that Apple negotiated a special deal with the Irish tax authorities.[9] If this special deal is not already known by finance ministers around the world, it will be interesting to see their reaction when it becomes known after this hearing.[10]

  o Very little income recorded in countries other than the US and Ireland - Although 60% of its 2011 sales were to customers in countries other than the US and Ireland, only 6% of the consolidated pre-tax income was recorded in such countries. This result was accomplished by recording substantially all of the pre-tax income from customers outside of the Americas in ASI. Entities in foreign countries other than Ireland received only a small commission for the sale of goods into their respective countries.

  o Irish holding company managed and controlled outside of Ireland - Apple's foreign holding company, Apple Operations International (AOI) is incorporated in Ireland but managed and controlled elsewhere.[11] As a result, dividends received by AOI from both Irish and non-Irish companies escape Irish taxation. If AOI was taxed in Ireland like other Irish corporations, it would have incurred a 25% tax on dividends received from subsidiaries located in non-EU countries.[12]

- Apple avoided substantial US taxation by side-stepping Subpart F income - Subpart F of the US tax law was designed to tax passive income of the type generated by Apple's Irish operations. However, Apple avoided substantial subpart F income[13] through use of (i) the check-boxregulations, (ii) CFC look-through rules, and (iii) the same country exception. In the future, Apple may also attempt to argue the contract manufacturing exemption applies. See Section 1 of the Appendix accompanying this testimony for a discussion of how these techniques are used by US MNCs to avoid Subpart income.

- US operations are less profitable than non-US operations - The ratio of pre-tax income to net sales is 24% in the US, but 36% for non-US operations. In addition, both general and administrative (G&A) expenses and sales, marketing, and development (SM&D) expenses as shown on Apple's 2011 consolidating income statement appear to be disproportionately allocated to the US.

  If G&A expenses were allocated based on pre-tax income and SM&D expenses were allocated on the basis of sales, US pre-tax income would increase by approximately $2.2 billion and the ratios of pre-tax income to net sales would become 30% in the US and 33% outside the US. The resulting 3% difference in profitability (as opposed to the 12% actual difference) could be easily explained by differences in product mix around the world.

  Thus, even after taking into account the cost sharing arrangement, Apple's allocations of G&A and SM&D expenses between US and non-US operations are curious and could contribute to the decreased profitability of US vs. non-US operations. The impact of these allocations on the 2011 US tax liability could be as much as $2.2 billion x 35% = $0.8 billion.

In summary, by entering into the cost sharing agreement with its Ireland affiliates and negotiating a special tax deal with Ireland, Apple was able to shift approximately $22 billion of its 2011 pre-tax income out of the US into Ireland and incur an immaterial amount of Irish tax. If such income had been taxable in the US, Apple would have incurred approximately $22 billion x 35% = $7.7 billion of additional US federal tax.[14] As demonstrated by many prior studies, Apple's efforts to shift income from the US to a tax haven jurisdiction are not unique. However, given Apple's overall profitability the magnitude of income shifting is startling.

KEY STEPS TO SHIFTING INCOME OVERSEAS

Before discussing the tax policy implications of income shifting, it may be helpful to summarize the key steps US MNCs take to shift income to tax haven jurisdictions and ultimately obtain a financial statement tax benefit. It is important to note that the goal of international tax planners is to shift income with minimum disruption to the business's operations. Thus, the transfer of intangible assets (or the use of creative financing structures) is clearly preferred since it involves only minimal disruption to a MNC's normal operations.[15]

- *Contribute equity to a foreign subsidiary* - An equity contribution to a foreign subsidiary is usually the first step in shifting income out of the US. For example, if a US parent contributes $1 billion of cash to a tax haven affiliate and the tax haven affiliate invests the $1 billion at a 10% rate of return, the US parent has effectively shifted $1 billion x 10% = $100 million of income out of the US *annually.* Although annually shifting $100 million of income can produce a significant tax benefit, US MNCs can often shift further income as described below.

  Transfer valuable intangible asset, but minimize the compensation paid - A transfer may be accomplished through a variety of means (e.g., a cost sharing arrangement with or without a buy-in payment, an outright sale, a license, or a contribution to capital). More importantly, since the valuation of unique intangible assets is extraordinarily difficult, MNCs have a significant incentive to assign the lowest possible value. The use of a cost sharing arrangement allows a US MNC to shift an intangible asset before it is fully developed and therefore assign an even lower value. And in some cases (e.g., Apple), the transfer takes place before any material development in which case the tax haven affiliate only needs to share costs with its US parent and does not need to make a buy-in payment.

  Continuing with the prior example, assume the tax haven affiliate uses the $1 billion contributed from its US parent to acquire an intangible asset with an estimated value of somewhere between $1 billion and $5 billion. In such case, if the tax haven affiliate acquires the intangible asset for $1 billion, but it ultimately turns out to be worth $5 billion, the US MNC has effectively shifted another $4 billion of income out of the US (i.e., in addition to the earnings on the original $1 billion equity contribution).

  It should be noted that even if the payment from the tax haven affiliate to the US parent is at true fair market value for the intangible assets transferred, as described above the US parent has effectively shifted income to the tax haven affiliate by virtue of the equity contribution.

- *Isolate substantial non-US income in the tax haven entity* - This can be accomplished by a variety of means. In Apple's case, substantially all of the non-US income was isolated in Ireland. This result was achieved by having ASI own the development rights outside of the

Americas coupled with ASI entering into a contract manufacturing agreement with a 3rd party supplier located in China (i.e., Foxconn). ASI then sold the products manufactured by Foxconn to other Apple entities. ASI was treated as the principal in the transaction and Apple's other foreign entities appear to have only received relatively small commissions for aiding in the distribution and sale of Apple products to customers. The end result was that substantially all of the income was isolated in ASI.

In tax planning structures used by other MNCs, the valuable intangible may be held in a separate tax haven entity that charges a substantial royalty to operating entities. Regardless of the actual structure used, the goal is isolate as much non-US income as possible in tax haven entities.

- *Avoid Subpart F income for US tax purposes* - As described previously, Subpart F income was originally designed to tax passive income (e.g., interest, dividends, and income related to intangible assets) earned by foreign affiliates of US MNCs. However, as described in Section 1 of the Appendix to this testimony various techniques can be used to avoid subpart F income (e.g., check-the-box regulations, CFC look-through rule, manufacturing exemption, and to a lesser extent the same country exception).

- Adopt "indefinite reinvestment" assumption for financial accounting purposes – Even though a US MNC may successfully shift income to a tax haven entity, current US tax law still imposes a tax upon repatriation of the earnings from the foreign subsidiary to the US. Accounting rules generally require that a deferred US tax expense be recorded for the potential future US tax upon repatriation. However, this deferred tax expense is not recorded if the earnings of the foreign subsidiary are considered "indefinitely reinvested". Most US MNCs assume 100% or a very high percentage of their earnings in tax haven affiliates are indefinitely reinvested. Apple is relatively conservative and only assumes approximately 50% of the earnings from their Irish affiliates are indefinitely reinvested.[16]

The above discussion focused primarily on the transfer of intangible assets, but US MNCs also use various creative financing structures to shift income to tax haven affiliates (for an example, see Section 1.2 of the Appendix to this testimony). In addition, as possibly demonstrated by Apple's disproportionate allocation of G&A and SM&D costs to the US, routine cost allocations can also be a method used to shift income out of the US. Although the amount of income transferred overseas from the use of creative financing structures and routine cost allocations can be very material, the income transferred from intangible assets is usually even more material.

KEY TAX POLICY QUESTIONS

Given this background, the question quickly becomes: What action, if any, should Congress take to address income shifting? Unfortunately, there is no silver bullet. Plus, the specific action may depend upon Congress's view on the following questions:

- Are US policy makers only concerned about US MNCs competing against foreign MNCs, or are they also concerned about the ability of US domestic businesses to compete with US MNCs that can shift substantial income offshore?
- Does current US tax law really put US MNCs at a competitive disadvantage vs. foreign MNCs? Should the US act unilaterally, or wait for OECD or some other global action?
- If the US acts and cracks down too much on US MNCs, could US MNCs eventually shift even more operations overseas or expatriate their headquarters from the US?
- Assuming global consensus is not forthcoming, is there a US policy response that could balance the competing policy goals?
- Should the US retain the arms-length standard and if so, what protections are needed to protect the corporate tax base from erosion?

These are all hard questions that could cause reasonable policy makers to disagree. Each is discussed in more detail below.

- *US MNCs vs. US domestic companies* - US MNCs have done an excellent job of framing the competitiveness issue in terms of US MNCs competing against foreign MNCs. However, that is only half of the competitive issue. If US MNCs are able to shift substantial income offshore, US domestic companies could be put at a competitive disadvantage. In addition, in order to compete, US domestic companies may decide they need to move some of their operations offshore with the resulting loss in jobs and US taxable income.

Given US domestic companies currently employ all of their workers in the US, putting them at a disadvantage may not be the best answer for a country that is struggling with persistently high unemployment. Nevertheless, by the same token, US tax policy should attempt to avoid putting US MNCs at a competitive disadvantage vs. foreign MNCs. This leads to the second key question.

- *Are US MNC's disadvantaged?* - Again, US MNCs prefer to focus on the element of US tax law that is competitively detrimental; the so-called lockout effect resulting from the taxation of dividends repatriated to the US. The lockout effect is a real problem for US MNCs, but one needs to also focus on what is causing the problem. Specifically, I believe the US MNC's lockout problem is primarily driven by the excessive amounts of income they have shifted outside the US. If such income had not been shifted, US MNCs would likely have a substantially smaller or non-existent lockout issue. Said differently, many US MNCs have been "hoisted on their own petard" by virtue of excessive income shifting out of the US.

Often forgotten in the discussion are the elements of US tax law that may give US MNCs a competitive advantage over foreign MNCs, including:

- o Subpart F rules are no longer effective - Historically the US had the toughest rules surrounding passive income earned by foreign subsidiaries, but with the introduction of the check-the-box regulations, the CFC look-through, and the contract manufacturing exemption the US rules are no longer effective. As a practical matter US MNCs can easily avoid the rules and could be at a competitive advantage vs. MNCs from certain countries.

- o Ability to obtain a US deduction for expenses related to foreign subsidiaries – This is especially the case for interest expense incurred in the US, but is also applicable to other expenses. For example, under current US tax law a US MNC can borrow in the US and fund its foreign operating entities through various creative tax structures.17 The end result is the US MNC claims a US tax deduction for interest related to foreign operations, but can avoid the recognition of any foreign income.

- o Cross-crediting of Foreign Tax Credits (FTCs) – If a foreign subsidiary incurs a relatively high tax rate (e.g., through tax losses not being allowed to be carried back to prior years), US tax law allows those high taxes to offset the US tax that would otherwise be incurred on low-taxed foreign earnings. This benefit is generally not available to foreign MNCs.

Overall I believe US MNCs are generally no worse off than foreign MNCs and in many cases may be better off. As a result, if Congress decides to impose restrictions on transferring income overseas, US MNCs should not be put at a competitive disadvantage vs. foreign MNCs.

- *Unilateral vs. global action* - If a global consensus is possible within a reasonable period of time and would be effective, it would be the best way forward. Given the OECD is working feverishly on its Base Erosion and Profit Splitting project[18] and is scheduled to disclose the results in June or July of this year, it is possible that a global consensus could emerge. Thus, Congress should clearly keep abreast of future OECD recommendations. But if history is any guide, it often takes OECD recommendations many years to come to fruition.

In addition, it often takes leadership from the US or other countries to jump-start global consensus on an issue. For example, the US adoption of FATCA addressing the reporting of offshore accounts has resulted in significant global coordination and action in the past 2 years. Thus, there may be a need for the US to take action on income shifting to spur the rest of the world into action.

- *Could the US crack down too hard on US MNCs?* - The short answer is yes. Since there will always be countries that will offer MNCs an attractive location to operate and/or relocate their headquarters operations,[19] US policymakers need to be concerned about the long-term impact of any proposals. For example, if the US were to unilaterally adopt full worldwide taxation without deferral of active income from foreign subsidiaries, there would be significant risk that over time US MNCs would figure out a way to eventually expatriate out of the US to take advantage of substantially lower rates in tax havens.[20]

- *Assuming global consensus is not forthcoming, is there a US policy response that could balance the competing policy goals?* - Possibly, but it is very unlikely to get serious political consideration. One policy response that would allow US MNCs to compete effectively with foreign MNCs and not have a competitive advantage over US domestic businesses would be to lower the US corporate income tax rate to 15% or less and replace the lost revenue with another revenue source (e.g., VAT). Although they are collected differently, a VAT and the corporate income tax have some similarities. For example, when compared with a corporate income tax, a VAT does not allow a deduction for labor costs, but does allow a 100% deduction for capital expenditures.

- *Should the US retain the arms-length standard and if so, what protections are needed to protect the corporate tax base from erosion?* – These two questions have been at the heart of much of the recent debate surrounding international tax reform and will likely be the subject of much discussion by the OECD as it develops its recommendations for its BEPS project. My views on these two questions, in reverse order, are as follows:
  - o *Arms-length standard requires strong base erosion protections* - If the arms-length standard is retained and US MNCs can continue to (i) make equity contributions to foreign subsidiaries and (ii) transfer valuable intangible assets to foreign subsidiaries, it will be *crucial that* steps be taken to minimize base erosion (e.g., restoring the vitality of Subpart F, imposing a minimum tax on tax haven income, and limiting the deductibility of interest).
  - o *Abandoning the arms-length standard requires a determination of how income should be allocated* - For example, where should the income attributable to intangible assets developed in the US be taxed? If it is the US, will it cause US MNCs over time to transfer their research activities overseas? In summary, if the arms-length standard is abandoned, there will be a need to determine what factors of production should be used to allocate income to various tax jurisdictions. A multilateral approach would be strongly preferred, but if history is any guide, multilateral action is unlikely.

IMMEDIATE TAX POLICY RECOMMENDATIONS

Clearly there are many potential views on the key questions discussed above, and as a result there will be significant tax policy debates surrounding the appropriate taxation of MNCs. Nevertheless, given that it is generally agreed that the Subpart F rules have recently been "gutted", I believe Congress should seriously consider the following *sooner rather than later:*

- *Substantially restrict the check-the-box regulations, the CFC look-through rule, and the contract manufacturing exemption* – The Subpart F rules were designed to make it very difficult for passive income related to intangible assets and creative financing structures to be shifted out of the US. As discussed further in Section 1 of the Appendix of this testimony, the relatively recent adoption of the check-the-box regulations, the CFC look-through rule, and the contract manufacturing exemption have allowed US MNCs to effectively avoid the Subpart F rules.

Said differently, US MNCs have been able to shift income from *both*[21] the US and high-tax foreign countries and locate the income in a tax haven without much fear of triggering the US Subpart F rules. The following suggestions would help restore the original vitality of the Subpart F rules:

- o *Restrict check-the-box regulations for foreign corporations* – Although there could be several options to accomplish this goal, one is to require conformity in treatment between US and foreign tax law. For example, if the foreign entity is treated as a corporation for local tax law purposes, it should be treated as a corporation for US tax purposes. Another option would be to expand the list of per-se foreign corporations (i.e., foreign corporations treated as corporations for US tax law). In summary, the goal of any change to the check-the-box regulations should be to minimize the ability of MNCs to create hybrid entities whereby the entity is respected for one country and disregarded for the other.

- o *CFC look-through rule should at most apply to only dividends* - When the CFC look-through rule was enacted in 2006 I was personally stunned it was made applicable to payments other than dividends (e.g., interest and royalties). The end result has been that US MNCs can locate intangible assets and financing operations in tax havens and avoid Subpart F income. Congress should consider either totally eliminating the CFC look-through rule or alternatively only allow it to be applied to dividends.

- o Contract manufacturing exemption should be eliminated or substantially tightened – The original Subpart F rules were designed to exclude from US taxation the income earned by a foreign corporation to the extent (i) the property was manufactured in the foreign country or (ii) the property was sold to customers in such country. In the good old days, "manufactured" meant that the foreign corporation had a plant in the foreign country and actually manufactured something in the plant. Not so any more.

In 2008 the IRS and Treasury issued regulations allowing supervision of contract manufacturing to qualify as manufacturing.[22] As a result, the manufacturing exemption to Subpart F income has been greatly expanded allowing US MNCs to avoid substantial amounts of US taxable income.[23]

Although there could be many potential changes to the contract manufacturing exemption, the easiest solution would be to just eliminate it. Another option might be to only allow the foreign corporation to avoid Subpart F to the extent of the labor cost of the supervisory services provided plus some reasonable profit margin.

o   Same country exception should be modified - See discussion in Section 1.4 of the Appendix to this testimony.

It needs to be emphasized that all of these suggestions should be adopted as a package. If only one or two are adopted, US MNCs will be able to use the remaining techniques to accomplish their tax planning goals.

- *Increased transparency* – Currently it is very difficult for the IRS and tax administrators around the world to get a true picture of a US MNC's effort to shift income to low tax jurisdictions. As discussed in more detail in Section 2 of the Appendix to this testimony, there should be increased transparency surrounding the worldwide tax position of MNCs. Information might include a schedule summarizing where income is recorded for both financial accounting purposes and tax purposes, the amount of tax paid, and other information of potential use to tax administrators.

BROADER TAX REFORM RECOMMENDATIONS

Although the above proposals would be very beneficial in turning the clock back on income shifting by restoring the vitality of the Subpart F rules, additional tax policy changes will also be needed. The reason is that as long as the US (and the rest of the world) applies an arms-length pricing standard to transactions between controlled parties, there will always be opportunity for MNCs to shift income to tax havens. The arms-length standard was developed almost 100 years ago and from a pure conceptual basis it makes some sense.

The problem is getting it to work in practice, especially in today's world where:

- Production and distribution functions are no longer vertically integrated in one foreign country, MNCs exercise substantial or complete control over their foreign subsidiaries, Intangible assets being transferred are extraordinarily unique and difficult to value, and
- MNCs seek to exploit the arms-length standard by spending significant time and money developing plans to shift income to tax haven jurisdictions.

As a result, tax administrators around the world are wrestling with the issue of how to address the shifting of income by MNCs to tax haven jurisdictions. As I have stated previously, even though the IRS has greatly increased its resources for auditing transfer pricing issues, "anyone who believes the IRS can effectively enforce the arms-length standard is either eternally optimistic -- or delusional".[24]

For these reasons, a longer-term solution is ultimately needed. But again, there are many options [25] and the potential for reasonable policymakers to disagree. My personal recommendations, in order of preference, are below:

- *Obtain global consensus* - If a global consensus could be reached that little or no income should be allocated to tax havens, it would be a giant step forward. There would still need to be a determination as to how to allocate income between source and residence countries, but it would be very beneficial if tax havens could be substantially taken out of the equation.

How might this work? The scenario easiest to conceptualize would be a global agreement on some sort of formula apportionment method, but there are many others. For example, source countries could be required to impose withholding taxes on payments to a tax haven jurisdiction. In addition, a headquarters/residence country could be required to impose a tax on all income earned by foreign subsidiaries located in tax havens.[26]

Unfortunately, I am not optimistic about the chances of an effective global agreement occurring anytime soon, but if it does happen, it would be welcomed. Given global agreement may not occur or may not be effective, US policymakers should also consider the unilateral options discussed below.

- *Lower the corporate income tax to no more than 15%[27] and adopt some other revenue source (e.g., VAT, energy taxes, and/or financial transactions taxes*) – In a world where many foreign countries are competing through low corporate income tax rates, one has to wonder whether the US will ultimately have to capitulate and join the fray. A corporate income tax in the 15% range could balance the competitive issues facing both US MNCs and domestic corporations. Given the political issues faced by such a proposal, however, the chances of this policy suggestion being adopted any time soon are not very high. Therefore, I will say no more.

- Keep the arms-length standard, but...
  - ○ Overlay a minimum tax on tax haven earnings – There are many different variations to this approach. In determining which to adopt, Congress should prefer those options that are easiest to administer. One general class of options is to identify a low-tax country (or foreign corporation) and apply a tax to some or all of the income from such country (or foreign corporation) without the benefit of a foreign tax credit.
  - ○ If an approach of this type is ultimately adopted, I strongly recommend that all earnings of a tax haven should be included as opposed to trying to determine (i) excess earnings, or (ii) earnings attributable to intangible assets. This would rule out many proposals, including options "A" and "C" of the House Ways and Means October 2011 proposals.[28]
  - ○ Again, from a simplicity perspective, it would be preferable if it is clear whether a foreign country is, or is not, a tax haven. Unfortunately, given the special deals that country's like Ireland are willing to make with MNCs like Apple, relying on a published corporate tax rate may not work.[29] Thus, one may need to focus on a specific company's fact pattern in the country.
  - ○ Disallow US deductions for expenses attributable to foreign income – Currently US tax law allows a MNC to take deductions (e.g., interest and G&A) in the US for expenses that may be attributable to foreign subsidiaries. This is especially a problem for those US MNCs that incur interest expense in the US and then equity fund foreign subsidiaries.[30] US tax law should not allow these deductions until the foreign income is recognized in the US.
- *Unilaterally replace the arms-length standard with a formula apportionment approach* – This approach has been advocated by some31 and in my opinion is better than current law. However, there are potential issues that would need to be addressed, including (i)

whether to base the formula on sales, or sales plus other factors of production (e.g., employees and/or tangible assets), and (ii) the need for anti-abuse rules in cases where sales are made to an intermediary in a tax haven country.

It should be emphasized the above discussion is equally applicable whether Congress decides to continue with the current hybrid system of worldwide taxation, or adopts a territorial system. However, if the US wants to adopt a territorial system, it should only be adopted if there is a high degree of confidence that the risk of income shifting is minimal. The least desirable option is to keep the current US tax system for taxing MNCs without any changes. The reason for this being that the current Subpart F rules effectively allow US MNCs to shift substantial amounts of income out of the US.

* * *

This concludes my testimony and I would be pleased to answer any questions.

APPENDIX

1. Techniques for Avoiding Subpart F Income[32]

Once a US MNC has successfully shifted income into a tax haven, it must attempt to avoid the inclusion of such income in its US tax return (i.e., avoid Subpart F income). Subpart F is a very complicated area of the tax law, and the discussion below briefly discusses only two items.

- *Foreign Base Company Sales Income (FBCSI)[33]* – This provision is designed to tax income earned by a foreign subsidiary when the subsidiary does not materially participate in the generation of the income *and* the subsidiary either buys or sells personal property from or to a related party.

  For example, if an Irish subsidiary of Apple (e.g., ASI) purchases personal property from a Chinese supplier (e.g. Foxconn) and sells the property to another Apple subsidiary outside of Ireland that in turns sells to a 3rd party customer, the FBSCI rules are generally intended to apply to both ASI and the related party. However, as described below,

Apple is able to avoid the FBCSI rules through the use of the "check-the-box" regulations,[34] and possibly in the future through the so-called "contract manufacturing" exemption.[35]

- *Foreign Personal Holding Company Income (FPHCI)[36]* – This provision is designed to tax interest, dividends, royalties, and certain other types of passive income earned by a foreign subsidiary. Thus, if a foreign subsidiary of Apple (e.g., AOI) directly or indirectly receives dividends or interest income from another Apple foreign subsidiary (e.g., AOE or ASI), the general FPHCI rules would treat such income as taxable in the US. However, as described below, Apple is able to avoid the FPHCI rules by use of (i) the check-the box regulations, (ii) the CFC look-through rule,[37] and/or (iii) the same country exception.[38]

In summary, Apple substantially avoids the application of these two subpart F provisions (i.e., FBCSI and FPHCI) through a combination of techniques referred to above (i.e., the check-the box regulations, the CFC look-through rule, the same-country exception). In addition, in the future Apple may be able to use the contract manufacturing exemption. See Sections 1.1 to 1.4 of this Appendix for more discussion of these techniques and specifically how Apple or other MNCs may use them.

1.1. Check-the-Box Regulations[39]

These regulations adopted by the IRS/Treasury in 1996 allow US MNCs to create so-called "hybrid entities" where the entities may be taxed as an entity in one tax jurisdiction and either a pass-through entity or a disregarded entity in the other tax jurisdiction. Although there are multiple tax planning uses for hybrid entities, one of the most common is to treat an entity for US tax purposes as disregarded and therefore also disregard transactions between the entity and its parent.

Apple appears to benefit from the check-the-box regulations by treating many entities as disregarded for US tax purposes and thus transactions between such entities are disregarded. As a result, the income that could otherwise be taxable as Foreign Base Company Sales Income (FBCSI) or Foreign Personal Holding Company Income (FPHCI) in the US disappears.

Below is a simplified illustration of Apple's legal structure for its European sales:[40]

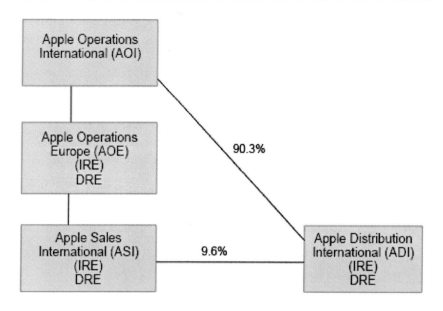

Notes:

1. AOI is incorporated in Ireland, but "managed and controlled" outside of Ireland. Therefore this entity is not currently taxed in any country.
2. Baldwin Holdings Unlimited, a British Virgin Islands entity, owns less than 0.1% of AOI, AOE, ASI, and ADI
3. IRE = Incorporated in Ireland. Note that AOI is not subject to tax in Ireland and ASI is only taxed on very limited activities.
4. DRE = disregarded entity for US tax purposes

From an operating perspective, ASI owns the right to Apple's development rights outside of the Americas. Thus, ASI contracts with Foxconn to manufacturer Apple products and immediately sells them to ADI who in turns sells the products further down the distribution chain.[41] Substantial pre-tax profits (e.g., $22 billion in fiscal 2011 and $74 billion for the 4 years 2009 to 2012) are accumulated in ASI and relatively minor amounts of pre-tax profit are reported in ADI and other downstream affiliates.

From a legal perspective, AOI, AOE, ASI, and ADI are all incorporated in Ireland and treated as corporations under Irish law.[42] However, from a US tax perspective, Apple has made check-the-box elections on AOE, ASI, ADI, and other Apple affiliates further down the distribution chain. The end result is that

Apple appears to treat AOI, AOE, ASI, ADI, and the downstream affiliates as one big entity for US tax purposes and therefore the inter-entity transactions are ignored.[43]

For example, if sales from ASI to ADI were respected (i.e., not ignored), Apple could have FBCSI.[44] However, because Apple has checked-the-box on ADI to treat ADI as a disregarded entity, sales between ASI and ADI are totally ignored for US tax purposes and FBCSI is avoided.

In addition, ASI makes substantial annual dividends (e.g., over $6 billion in fiscal 2011) to AOE that in turn makes dividends to its parent, AOI. Although dividends are a class of income that can cause FPHCI, Apple avoids the issue by checking-the-box on both AOE and ASI to treat them as disregarded entities. As a result the dividends from ASI to AOE and AOE to AOI are disregarded and Apple avoids FPHCI.[45]

1.2. CFC Look-Through Rule[46]

The CFC look-through rule was enacted in 2006 by Congress to allow US MNCs to re-characterize what would otherwise be subpart F income (e.g., dividends, interest, and royalties) by looking-through to the character of the income earned by the entity paying the dividend, interest, royalty, etc...

For example, when AOI receives dividends from both Irish entities (e.g., AOE) and non-Irish entities, such dividends could be FPHCI subject to US tax. However, to the extent the dividends are attributable to active income from the subsidiary, the CFC look-through rule effectively re-characterizes the dividend income as active income and FPHCI can be avoided.

Apple appears to have benefited from the application of the CFC look-through rule to re-characterize dividend income as active income and therefore avoid Subpart F income. It does not appear that any substantial amount of interest or royalty income was re-characterized as operating income. Nevertheless, it is important to illustrate how the CFC look-through rule can be used to avoid US taxable income in certain cases that many may view as abusive.

- *Baseline Case* - As a baseline, assume a US Parent borrows from a 3rd party in the US and on-lends the borrowed funds to foreign operating subsidiaries located in a relatively high tax country. For US tax purposes the US Parent will have interest expense from the 3rd party and interest income from the foreign operating subsidiaries. Presumably the two amounts will roughly offset one another and thus there is no US tax consequence from the US Parent acting as an intermediary for the loan. For foreign tax purposes, the foreign operating subsidiaries should receive a tax deduction for the interest paid to the US parent. The diagram below illustrates this simple funding scenario.

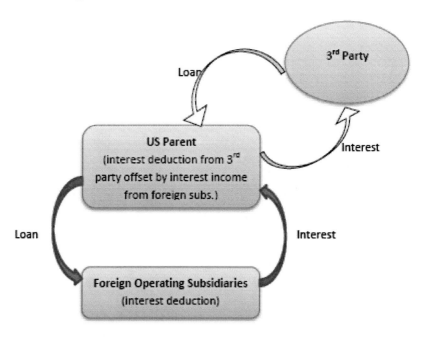

In summary, the tax results from this simple baseline case seem to be very reasonable (i.e., the interest deduction is ultimately claimed in the foreign operating subsidiaries and there is no material deduction or income in the US parent).

- *Alternative Scenario* – Assume the US Parent again borrows from a 3rd party, but instead of the US Parent directly on-lending to its foreign operating subsidiaries, the US Parent contributes the borrowed funds to the capital of a foreign subsidiary located in a country with no income tax (Tax Haven Subsidiary). Then assume the Tax Haven Subsidiary loans the funds to foreign operating subsidiaries around the world in high-tax countries. The diagram below illustrates this more complicated funding structure.

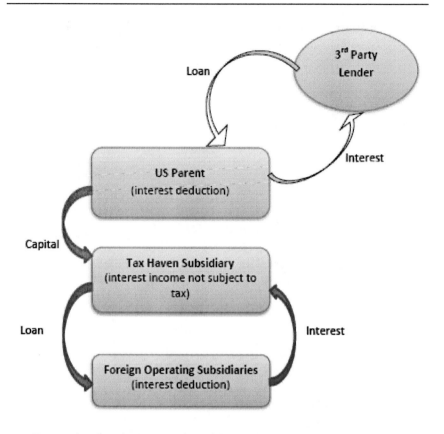

From a legal entity perspective, the consequences of this structure are (i) the US Parent will have interest expense, (ii) the Tax Haven Subsidiary will have interest income, and (iii) the Foreign Operating Subsidiaries will have interest expense. From a US tax perspective, the question is whether the interest income earned by the Tax Haven subsidiary is FPHCI and therefore included in the US Parent's US taxable income?

The answer is the Subpart F rules were originally designed to significantly discourage this sort of funding structure. However, after application of the CFC look-through rule, the interest income earned by the Tax Haven Subsidiary will likely be re-characterized as operating income because the interest is being paid from an operating entity. The end result is that the US MNC will obtain two interest deductions (i.e., one at the US Parent and one at the Foreign Operating Subsidiaries) and not pay tax on the interest income in any location. When compared with the Baseline Case, this Alternative Scenario results in income being excluded from the US tax return.

Before discussing other techniques for avoiding Subpart F income, two additional items should be noted about the above financing structure:

- *Check-the-box regulations can accomplish the same result* - If the CFC look-through rule did not exist, the US MNC in the example above could accomplish the same result by checking-the-box on foreign operating subsidiaries to treat them as disregarded entities. Since the foreign operating subsidiaries would be viewed as part of the Tax Haven Subsidiary, the interest income would be disregarded for US tax purposes (i.e., it disappears).
- *Structure is also applicable to intangible assets* - The example above assumes the Tax Haven Subsidiary was used as a finance vehicle. However, the above simple structure can also be used to avoid FPHCI on royalties from intangible assets. For example, a Tax Haven Subsidiary could use funds contributed by its US Parent to acquire rights to intangible assets that are then licensed (or sublicensed) to foreign operating subsidiaries around the world. In summary, FPHCI on the royalty income can be avoided through either (i) the application of the CFC look-through rule, or (ii) the check-the-box regulations.

1.3. Contract Manufacturing Exemption[47]

The original Subpart F rules were designed to exclude from US taxation the income earned by a foreign corporation to the extent (i) the property was manufactured in the foreign country or (ii) the property was sold to customers in such country. In the good old days, "manufactured" meant that the foreign corporation had a plant in the foreign country and actually manufactured something in the plant. Not so any more.

With the advent of contract manufacturing, manufacturing is now often done by a third party in a low-cost country (e.g., China, Philippines, and Bangladesh). Thus, the tax issue became whether a foreign subsidiary of a US MNC could qualify for the manufacturing exemption to Subpart F by supervising or making a "substantial contribution" to the contract manufacturing operations of the 3rd party.

In 2008 the IRS and Treasury issued regulations allowing supervision of contract manufacturing to qualify as a "substantial contribution" to the manufacturing process. As a result, foreign corporations may qualify for the contract manufacturing exemption to Subpart F and effectively avoid substantial amounts of US taxable income.[48]

Although there could be many potential changes to the contract manufacturing exemption, the easiest would be to just eliminate it. Another option might be to only allow the foreign corporation to avoid Subpart F to the extent of the labor cost of such supervisory services plus some reasonable profit margin.

1.4. Same-Country Exception[49]

Although the check-the-box regulations, the CFC look-through rule, and the contract manufacturing exemption are the primary tools US MNCs use to avoid Subpart F income, one other tool worth mentioning surrounds the same-country exception. This is especially true in Apple's case because their fact pattern would allow them to benefit from the same-country exception if for some reason the three primary tools for avoiding Subpart F were not available.

As previously described in this testimony, AOI is an entity incorporated in Ireland but considered managed and controlled elsewhere to avoid Irish tax on dividends from non-Irish companies. In 2011 there were over $6 billion of dividends from AOE to AOI not taxed in Ireland. From a US tax perspective, the question is whether the $6 billion of dividends from AOE to AOI should be considered FPHCI and therefore immediately taxed in the US? Thanks to the application of the check-the-box regulations and/or the CFC look-through rules; this $6 billion is not FPHCI. However, it should be noted that even if the check-the-box regulations and the CFC look-through rule were unavailable, it appears Apple would not have FPHCI by virtue of the same-country exception. The same-country exception provides that dividends and interest received by AOI from AOE will not be FPHCI because AOE is created or organized in the same country as AOI (i.e., Ireland). This is the case even though AOI is not taxed in Ireland because it is "managed and controlled" outside of Ireland. Thus, one needs to question whether it is appropriate to allow the same-country exception in this type of case. Congress may want to consider modifying the same country exception to provide that it is only applicable if the two entities are subject to taxation in the same country.

2. PROPOSAL FOR INCREASED TRANSPARENCY

Information is money - - - Although this phrase is used commonly in the business world to refer to the ability of businesses to generate revenue from the possession of valuable information, it is also applicable to the relationship between the tax departments of many MNCs and tax officials around the world. Historically many MNCs have made it *very difficult* for tax officials to obtain information with the end result that either the tax officials (i) don't obtain the necessary information to propose potential audit adjustments, or (ii) spend so much time attempting to obtain information on one issue that they don't have as much time to investigate other issues.

Because of this concern, the IRS adopted Schedule UTP in 2010.[50] Schedule UTP now requires corporations to disclose tax issues to the IRS when the corporation has recorded a tax reserve for an issue on its audited financial statement. However, the disclosure on Schedule UTP is limited to a few sentences and is only intended to identify issues for the IRS on a very general level.

When Schedule UTP was being developed there was significant concern expressed by some at the IRS that additional information surrounding transfer pricing issues was needed. A decision was made that Schedule UTP was not the correct vehicle to request such information,[51] but this decision did not foreclose the possibility of future additional disclosure specifically targeting transfer pricing. Since I departed the IRS almost 3 years ago, I do not know whether the IRS is currently considering any additional transfer pricing disclosure. If not, they should, especially for publicly traded MNCs. Additional transfer pricing related information could help the IRS more quickly identify the location and scope of any income shifting.

Information could be obtained on an aggregate basis (e.g., US vs. non-US), country-by-country basis, and/or on an entity by entity basis (i.e., at the controlled foreign corporation (CFC) level). At a minimum, aggregate information could include the following:

	Pre-tax income	Tax Expense		Liability per the tax return	# of Employees	Sales	Other Information
		Current	Deferred				
US				n/a			
Non-US							
Consolidated financial statements							

Note that some of this information is now included in audited financial statements, but some is not. In addition to making this available to the IRS and other tax authorities, one could also consider making it available to the public.

Although MNCs would surely complain, the information disclosed is relatively aggregated and in many cases is already publicly disclosed.

Additional information could also be collected on a CFC by CFC basis[52], including that requested in the following two tables:

	How is the entity classified (i.e., taxable, disregarded, or flow-through)?		Financial Statement Information				
CFC name	Foreign Tax Purposes	US Tax Purposes	Pre-Tax Income	Federal tax expense		Foreign tax expense	
				Current	Deferred	Current	Deferred
A							
B							
etc...							

CFC name	US Federal tax		Foreign tax		Other Information (e.g., # of employees and/or employee compensation)
	E+P	Subpart f	Taxable income	Tax Liability	
A					
B					
etc...					

Although some of this information is currently collected on IRS Form *5471, Information Return of US Persons with Respect to Certain Foreign Corporations,* some is not. In addition, I suspect there is additional information that IRS field agents would find useful. My bottom line suggestion is the IRS should determine what information would be useful and design a form to collect it. Whether this is a new form, or some variation of an existing form does not matter. The key is to allow IRS agents to quickly identify where income is being shifted to low-tax jurisdictions, and how such shifting is being accomplished.

End Notes

[1] During the period 2009-2012, the total pre-tax income recorded in one Irish entity, Apple Sales International (ASI), was approximately $74 billion. There were no employees in ASI until 2012 when 250 employees appear to have been transferred from another Irish entity.

[2] At June 2011, Apple's total employees worldwide were approximately 59,000 while total worldwide compensation was approximately $6.9 billion.

[3] Customer location based upon information provided in the 2011 consolidated financial statements, except for Ireland that was assumed to be <1% based upon the relative size of Ireland's population.

[4] These observations are not meant to conclude that Apple has done anything improper from a tax planning perspective. Such determination could only be made after a detailed audit of Apple's facts.

[5] Technically the agreement is with both ASI and Apple Operations Europe (AOE). However, it appears that AOE is functioning as primarily a holding company.

[6] ASI's pre-tax income in 2102, 2010, and 2009 was approximately $36 billion, $12 billion, and $ 4 billion respectively. Thus, for the four-year period 2009 to 2012, Apple's cost sharing agreement with ASI effectively resulted in a reduction in US pre-tax income of approximately $74 billion. In addition, the cost sharing expenses incurred by Apple's Irish entities for the period 2004 to 2011 were approximately $5 billion. Thus, Apple's Irish entities were able to spend $5 billion to obtain pre-tax income before such expenses of at least $74 billion + $5 billion = $79 billion.

[7] 64% of the consolidated pre-tax income was recorded in Ireland when (i) only 4% of the global workforce and 3% of global compensation expense is located in Ireland, and (ii) approximately 1% of customers are located in Ireland.

[8] The total tax expense recorded in Apple's 2011 Consolidating Income Statement for Apple Sales International (ASI) and Apple Operations Europe (AOE) was only $13 million (i.e., an effective tax rate of $13 million/$22billion = 0.06%). Per information supplied by Apple, ASI is not taxed as an Irish resident corporation. Thus, it appears to only be taxed on certain business activity in Ireland.

[9] Apple reportedly confirmed this hypothesis in discussions with the PSI staff.

[10] During negotiations of Ireland's bailout by the EU in 2010, there was reportedly discussion about forcing Ireland to increase its corporate tax rate and eliminate special tax deals. Nevertheless, Ireland was able to obtain its bailout without any material tax changes. Will the EU be so generous in the future?

[11] Apple apparently has not affirmatively concluded where this entity is "managed and controlled", but it is worth noting that the US would seem to be the only other possibility. For example, Board of Directors meetings are held in the US and bank accounts are in the US.

[12] Since very little income was recorded in other foreign countries, this tax planning does not appear to have produced a material tax benefit to Apple. In addition, the 25% tax could have been reduced in certain cases through tax treaties.

[13] Apple did report some subpart F income, but the amount was relatively immaterial (i.e., approximately $100 million in 2011) and related to interest income from 3rd parties.

[14] There also could have been additional state and local taxes

[15] However, in order to provide some substance to the transactions for tax purposes, international tax planners will often move some assets or employees to improve the optics of the transfer. For example, Apple had approximately 2,500 employees in Ireland in 2011, but this was still only 4% of its 2011 global workforce with the end result that 60% of Apple's 2011 global pre-tax income was transferred to Ireland.

[16] This is one of the reasons why Apple's effective tax rate disclosed on its 2011 financial statements is 24.2%. If Apple had assumed that 100% of its foreign earnings were indefinitely reinvested, the disclosed 2011 effective tax rate would have decreased to approximately 12.8% (i.e., $8.283 billion of reported tax expense – $3.917 billion increase

in the deferred tax liability on unremitted foreign earnings = $4.366 billion adjusted tax expense/$34.205 billion of pre-tax income).

[17] See Section 1.2 of the Appendix of this testimony for a description of a specific financing structure.

[18] For more information see http://www.oecd.org/tax/beps.htm.

[19] In addition, countries that are not tax havens could still engage in tax competition (i.e., the UK and other countries continue to decrease their corporate tax rates).

[20] This would likely be the case even if the US decreased its corporate tax rate to somewhere in the 20-30% range. A US tax rate of 15% or less would likely be needed to minimize the incentive for US MNCs to expatriate if deferral were eliminated.

[21] This is an important point. US MNCs prefer to discuss the use of these techniques to strip income from high-tax foreign countries, rather than the US. However, in reality they are using these techniques to aid the shifting of income from both the US and high-tax foreign countries to tax havens.

[22] Regulation 1.954-3(a)(4)(iv).

[23] Said differently, the combination of the ability of US MNCs to transfer valuable intellectual property to a foreign corporation coupled with the foreign corporation's ability to then enter into a contract manufacturing arrangement allows for the shifting of significant income out of the US.

[24] See Tax Notes 121 (Jan. 2, 2012) available at http://papers.ssrn.com/sol3/papers.cfm?abstract id=1975975

[25] For example, see International Competitiveness: Senate Finance Committee Staff Tax Reform Options for Discussion (May 9, 2013) available at www.finance.senate.gov/issue/? id=0587e4b4-9f98-4a70-85b0- 0033c4f14883.

[26] A special rule would be needed to address MNCs headquartered or incorporated in tax havens.

[27] If the US corporate income tax rate is above 15% there is still will be a significant incentive for US MNCs to shift their income or operations to low-tax jurisdictions. As a result, even though discussions about lowering the US corporate income tax rate to the 25%-30% range are positive, I don't believe they will materially alter the incentive to shift income and operations out of the US.

[28] See http://waysandmeans.house.gov/uploadedfiles/discussiondraft.pdf

[29] One option could be to rely on the published tax law of a country, but make it clear that if any special deals are discovered the country will automatically be considered a tax haven (i.e., a death penalty).

[30] Or have much higher ratios of equity to assets.

[31] For example, Reuven Avi-Yonah and Michael Durst.

[32] IRC sections 951 to 965

[33] IRC 954(d)

[34] 301.7701-3. Apple reportedly claims they do not currently rely on the contract manufacturing exemption.

[35] 1.954-3(a)(4)(iv). Apple did not have any employees in ASI until 2012 and therefore was not eligible for the contract manufacturing exemption. One wonders whether in the future Apple will also argue that it avoids Subpart F by relying on the contract manufacturing exemption applies (i.e., have two arguments for avoiding Subpart F).

[36] IRC 954 (c).

[37] IRC 954(c)(6).

[38] IRC 954(c)(3).

[39] 301.7701-3.

[40] The information obtained is from a review of the information Apple supplied PSI.

[41] In some cases, it appears the products may be sold to 3rd parties while in other cases they are sold to Apple affiliates who eventually sell to 3rd parties.

[42] AOE, ASI, and ADI are all taxed in Ireland, but AOI is not because it is managed and controlled elsewhere. In addition, ASI after 2009 ASI reportedly does not meet Irish residency requirements and therefore is only taxed on certain limited business activity in Ireland.

[43] It is not clear how Apple treats Baldwin Holdings' <0.1% ownership in the disregarded entities. If Baldwin Holdings' ownership is respected, technically one would expect ASI, AOE, and ADI to be partnerships for US tax purposes which could add complications to the US tax analysis.

[44] As discussed in Section 1.3 of this Appendix, FBCSI can also be avoided if ASI is a substantial participant in the manufacturing process. Thus, the contract manufacturing rules may allow Apple to avoid FBCSI but it is not clear whether they are also making this argument.

[45] Apple may also be able to avoid FPHCI by virtue of either the CFC look-through-rules, or the same country exception.

[46] IRC 954(c)(6).

[47] Regulation 1.954-3(a)(4)(iv)

[48] Note that reportedly Apple does not currently rely on the contract manufacturing exemption. As Apple adds employees to ASI, it is possible that Apple could qualify in the future.

[49] IRC 954(c)(3)

[50] For information on Schedule UTP, see J. Richard (Dick) Harvey, Jr., Schedule UTP: An Insider's Summary of the Background, Key Concepts, and Major Issues, DePaul Business and Commerce Law Journal (Spring 2011) available at http://papers.ssrn.com/sol3/papers.cfm?abstract id=1782951.

[51] As the lead architect of Schedule UTP, I personally concurred with this decision.

[52] Because of its potential impact on a MNC's competitive position, I would not recommend publicly disclosing CFC specific information at this time. However, I would not completely rule it out at some time in the future (e.g., if MNCs continue to shift income to low-tax jurisdictions and other efforts have failed to prevent it).

In: Offshore Profit Shifting ...
Editor: Reny Toupin

ISBN: 978-1-62808-479-5
© 2013 Nova Science Publishers, Inc.

Chapter 4

TESTIMONY OF STEPHEN E. SHAY, PROFESSOR OF PRACTICE, HARVARD LAW SCHOOL. HEARING ON "OFFSHORE PROFIT SHIFTING AND THE U.S. TAX CODE – PART 2 (APPLE INC.)"[*]

Chairman Levin, Ranking Member McCain and members of the Subcommittee, thank you for the opportunity to testify on the important topic of shifting of profits offshore by U.S.

multinational corporations.[1] I am a Professor of Practice at Harvard Law School.[2] The views I am expressing are my personal views.

The Subcommittee and its staff should be commended for pursuing this important investigation. Protecting the existing U.S. tax base is an important responsibility of those in Congress and the Administration responsible for the fiscal health of the country. The revenue lost to tax base erosion and profit shifting is hard to estimate, but there is compelling evidence the amount lost is substantial. This revenue loss exacerbates the deficit and undermines public confidence in the tax system. Restoring revenue lost to base erosion and profit shifting would support investing in job-creating growth in the short term and reducing the deficit over the long term.

My testimony provides background information on the taxation of foreign income of U.S. multinationals earned through a controlled foreign corporation

---

[*] This is an edited, reformatted and augmented version of a testimony presented May 21, 2013 before the Senate Permanent Subcommittee on Investigations.

and on transfer pricing.[3] The testimony next discusses the information developed by the Subcommittee Staff regarding Apple's international tax planning and considers how current elements of U.S. tax law contribute to key elements of that planning. I will make a limited number of observations regarding the implications of the Subcommittee's Apple case study for tax law changes and conclude.

With the Chairman's permission, I would like submit my written testimony for the record and summarize my principal observations in oral remarks.

BACKGROUND: TAXATION OF FOREIGN SUBSIDIARY AND INCOME TRANSFER PRICING

Taxation of Foreign Subsidiary Income

Under current U.S. rules, a U.S. multinational is not taxed on active foreign income earned through a controlled foreign corporation (including, generally, a greater than 50% foreign subsidiary) until the earnings are distributed as a dividend.[4] This is commonly referred to as deferral.

The United States allows a domestic corporation that owns 10% or more of the voting stock of a foreign corporation a credit for foreign income taxes paid with respect to earnings received as a dividend in respect of that stock. A U.S. shareholder also may offset U.S. tax on a foreign dividend with excess foreign taxes paid in respect of other foreign income in the same foreign tax credit limitation category.[5] Accordingly, there is a residual U.S. tax on foreign earnings distributed as a dividend unless allowable foreign tax credits are sufficient to offset the U.S. tax. Interest expense and other deductions of a U.S. multinational, allocated to foreign income for purposes of determining the foreign tax credit limitation, are allowed as a current deduction even if the foreign income is deferred from current U.S. tax.

Through various devices, including gaps in anti-deferral provisions, many U.S. multinationals are able to reduce overall foreign taxes to burdens substantially below their effective U.S. tax rates. The combination of deferral of U.S. tax on foreign earnings, where the tax saved is reinvested at low foreign tax rates, and current deductions for expenses contributing to earning deferred income is a powerful incentive to shift income offshore. This incentive is magnified by financial accounting rules that allow undistributed foreign earnings to be taken into account in consolidated income without reserving for future U.S. tax if the earnings are considered indefinitely reinvested abroad.

Under the Internal Revenue Code's Subpart F anti-deferral rules, a United States shareholder in a controlled foreign corporation is subject to current income inclusion of its share of the controlled foreign corporation's "foreign personal holding company income," including interest, dividends, rents, royalties and capital gains not earned in an active business.[6] In addition to limiting deferral for passive income, certain other sales and services income earned through use of "base companies" may be currently included in a United States shareholder's income.[7] The two principal categories of active income that are subject to the anti-deferral rules are foreign base company sales income and foreign base company services income.[8] A United States shareholder may elect not to include currently Subpart F income that is subject to an effective rate of foreign tax greater than 90% of the highest U.S. corporate tax rate. The theory behind these base company sales and services provisions was that use of a base company in a lower-tax jurisdiction is an indicator of tax avoidance that should preclude the benefit of deferral. These provisions do not apply, however, to income earned in the country of organization of the corporation or to income from sales of property manufactured by the corporation.[9]

With the advent of U.S. "check-the-box" entity classification rules in 1997 and more recently the expansive acceptance of contract manufacturing by a third party for purposes of the "manufacturing" exception from foreign base company sales income, it is reasonably easy to avoid the reach of the Subpart F anti-deferral rules for a broad range of income. Statistics of Income data for 2006 show that approximately 80% of controlled foreign corporation earnings are retained and deferred from U.S. taxation, roughly 8% were distributed as dividends and 12% were currently taxed under Subpart F (and it should be recognized that Subpart F inclusions often are intentional in order to bring back earnings without triggering foreign withholding taxes).[10] For that year, the average effective rate of foreign tax on foreign earnings of controlled foreign corporations with positive foreign earnings was approximately 16.4%.[11]

The United States deferral system includes rules that restrict a controlled foreign corporation from making its offshore earnings available to its affiliated U.S. group other than through a taxable dividend distribution. The Section 956 "investment in U.S. property" rules, adopted in 1962 and frequently adjusted since, treat a controlled foreign corporation's offshore earnings that are invested in a broad range of U.S. investments, including a loan to its U.S. affiliates, as though the earnings were distributed as a dividend to a U.S. affiliate.[12] The investment in U.S. property rules include significant exceptions that are designed to allow investment of offshore earnings in U.S. portfolio securities.[13] The investment in U.S. property rules defend the residual U.S. tax on distributions but do not block holdings of U.S. portfolio investments.[14]

The effect of the investment in U.S. property rules, when they work properly, is to protect the U.S. income tax base by preventing a U.S. multinational from using earnings not taxed by the United States in its U.S. business.[15] These rules also restrict the advantage a U.S. multinational would have competing against a domestic U.S. business that will not have available low-taxed offshore earnings for use in its business. If there is leakage in the investment in U.S. property rules allowing deferred earnings to be loaned to the U.S. multinational's U.S. business without U.S. tax, the benefit of deferral on the earnings loaned would be preserved so financing from pre-U.S. tax earnings (after a foreign tax) would be available to the U.S. multinational but not its domestic competitors. The purpose of these rules is to prevent this, except in isolated cases of short-term loans.

Transfer Pricing

Transfer pricing generally refers to the prices charged between affiliates under common control for intercompany transactions, including sales or leases of tangible property, the performance of services and transfers by sale or license of intangible property rights. The transfer pricing rules of Section 482 attempt to ensure that taxpayers clearly reflect income attributable to controlled transactions and to prevent the avoidance of taxes with respect to such transactions.[16] The rules attempt to place a controlled taxpayer on tax parity with an uncontrolled taxpayer by determining the true taxable income of the controlled taxpayer.

From the first set of transfer pricing regulations in 1968, taxpayers have been permitted to share the costs of development of an intangible under a *bone fide* cost sharing arrangement as a means to determine which affiliates may earn returns attributable to the intangible. One of the substantial attractions for taxpayers of *bona fide* cost sharing is that the IRS generally will limit adjustments to the appropriate ratio for sharing costs. While the sharing ratio has been the subject of dispute, the far more substantial issue historically has been the valuation of contributions of pre-existing intangibles.[17]

If at the commencement of the cost sharing arrangement a participant possesses a resource, capability or right that is anticipated to contribute to development under the cost sharing arrangement, the other participants must compensate that participant for the fair market value of the contribution. The issue of pre-existing intangibles is referred to by practitioners as the "buy-in" problem, but the name is somewhat misleading. The "buy-in" concern is not limited to valuing intangible property that pre-exists the commencement of the cost sharing arrangement, but extends to the full range of contributions to development by affiliates whether or not they are participants in the arrangement. Paying for the full range and value of contributions has proved to be an Achilles heel (from the perspective of tax authorities) of cost sharing between related persons for tax purposes.

IRS and Treasury guidance regarding cost sharing has evolved through a series of developments reflecting successive problems with cost sharing in practice. The first limited guidance was given in final regulations in 1968. By the Tax Reform Act of 1986, it became clear that international transfer pricing was a substantial issue, particularly in relation to the territorial system adopted in Code Section 936 for Puerto Rico, so Section 482 was amended to permit a post-transfer review of the pricing of intangible property.[18] In 1988, the Treasury issued a White Paper on transfer pricing that sought to provide a sounder theoretical under pinning for the treatment of intangibles.[19] This was followed by 1992 proposed regulations that were heavily criticized by business and then 1995 final cost sharing regulations.

In 2007, the Treasury issued a report to Congress on transfer pricing that reported substantial evidence consistent with income shifting from non-arm's length pricing.[20] The 2007 Treasury Report acknowledged "that CSAs [cost sharing agreements] under the current regulations pose significant risk of income shifting from non-arm's length pricing." It reported on proposed regulations issued in 2005 that adopted a new "investor model" approach and that substantially expanded the newly-named "platform contributions" to the development of intangibles that should be compensated under new cost sharing arrangements. On the last day of 2008, the proposed regulations were largely adopted as temporary regulations, however, cost sharing agreements that were in existence on January 5, 2009 (and updated in certain respects), were subject to "grandfather" rules that insulated these agreements from the full force of the new rules. Final regulations were issued in 2011.[21]

The premise of the cost sharing rules is straightforward. If a participant shares all of the costs and all of the risks of developing a new intangible property, it is entitled like an entrepreneur to earn the returns from making that investment. As we were reminded in the global financial crisis of 2008, however, the application of theory and models in the messiness of the real world can lead to unintended or unanticipated results. As demonstrated by the repeated efforts to strengthen the cost sharing regulations and the continued evidence of income shifting to lower tax countries, the application of cost sharing in the context of the international taxation has proven to be highly problematic. This is in part because assumptions necessary for the theory of cost sharing to be valid, including that all contributions are fully accounted for, are nearly impossible to control in a real world setting.

The transfer pricing rules necessarily are an imprecise tool. The rules allow a taxpayer to fully comply by selecting the most advantageous price that falls within a range of allowable alternatives or, in respect of intangibles, by entering into a cost sharing arrangement.[22] The difficulties with administering transfer pricing rules in relation to a sophisticated multinational group are compounded where comparable third-party transactions are unavailable or inexact, as is the case with respect to most high-value intangible property, and by the flexibility afforded a multinational corporate group in planning and executing its global legal and pricing structure to minimize tax. The problems are exacerbated by the taxpayer's control over information and procedural advantages.[23]

The Subcommittee Staff's Apple Case Study: What Does It Tell Us?

The Apple information provided to the Subcommittee Staff offers visibility into the way one company organizes its affairs to shift very substantial amounts of income into low- or zero-tax jurisdictions. (Through its tax treatment of nonresident Irish corporations, Ireland may be considered both a low- and a zero-tax jurisdiction at the same time -- without explicitly providing a tax holiday.) The data developed by the Subcommittee staff supplements what is publicly available, but is limited to consolidating financial information (as opposed to tax return information) and written responses to Staff questions. Because of limitations on the information provided, and the circumstances under which it is made available,[24] the following discussion must be considered preliminary.

I take no position on the legal correctness or strength of any tax position taken by Apple. What are of interest are the techniques used to shift income to low-taxed countries and the scale of the income shifting that is possible.

Apple's business, organizational structure and international operations is described in the Staff Memorandum to the Subcommittee ("Staff Memorandum"). Apple is a remarkable and a remarkably successful company. In FY 2011, Apple had consolidated global revenues of $112 billion and earnings before tax of $34 billion.[25] Apple's FY 2011 global book tax rate was 24.2%, though Professor Harvey calculates it would be 12.8% if all of the Irish earnings are considered permanently reinvested.[26] Apple had approximately 59,000 employees worldwide in 2011.

Apple Transfer Pricing

The Apple companies in Ireland with respect to which information was provided (including companies organized in Ireland but reportedly tax resident nowhere) included two cost sharing participants under a longstanding cost sharing agreement with Apple for rights to sell products outside North America. In 2011, one of Apple's Irish cost sharing participants, Apple Sales International (ASI), contracted with third party manufacturers to make products and sold these products outside of North America. [27] Based on consolidating financials (without eliminations within those groups), in FY 2011 Apple's Irish companies earned approximately $22 billion in earnings before tax (EBT), or approximately 64% of global EBT. The Apple Irish companies' EBT to sales margin was 46% compared to 23% for Apple US.[28]

The effectiveness of Apple's transfer pricing and Irish nonresident company tax strategy is evident from the breakdown of Apple's FY 2011 EBT:

| | | FY2011 | | |
	US	Ireland	ROW	total
EBT ($ billions)	$10.2	$22.0	$2.0	$34
EBT share	30%	64%	6%	100%
Customers (approx.)	39%	1%	60%	100%

This illustrates in concrete terms for one company what has been shown in aggregate data, namely, that Apple aggressively shift earnings to a low- or zero-tax location.

To give a different measure, the Irish companies employed only 2,452 of Apple's 59,000 employees, yet they earned $22 billion in earnings before tax or over $9 *million* per employee. This actually is understated, since after the

2012 reorganization only 613 employees were assigned to the cost sharing companies (ASI and Apple Operations Europe). If 613 employees was the correct count for 2011, the EBT/employee would be *$35.8 million per employee* compared to an approximate average of $576 thousand per employee for all Apple employees.

The average effective book foreign tax rate for the Irish companies was under 1%. Apple described its low Irish tax rate as follows: "Since the early 1990s, the Government of Ireland has calculated Apple's taxable income in a way to produce an effective rate in low single digits ...since 2003 it has been 2% or less." According to Apple, the principal Irish companies in terms of income, Apple Operations Europe (AOE) and ASI, are not tax resident in Ireland. Based on Apple's disclosures so far, it is not clear that AOI, AOE and ASI are tax resident anywhere.

For U.S. tax purposes, Apple treated ASI and AOE as disregarded entities wholly-owned by Apple Operations International (AOI), an Irish-organized company with no employees or operations also considered by Apple to not be tax resident in Ireland. If the foregoing is correct, for U.S. tax purposes, all of the income earned by ASI and AOE is would be considered owned by AOI.

AOE and ASI, pay Irish tax only on their business carried on in Ireland. ASI is a party to the cost sharing agreement, but it is not clear where income attributable to the intangibles in which ASI has an interest is treated as earned; it appears to be allocated away from Ireland for Irish tax purposes, i.e., it could be what is fondly referred to by international tax planners as "ocean income." It would be difficult to achieve a less than 2% Irish effective tax rate if that income were subject to Irish tax at a 12.5% corporate tax rate (assuming it is considered trading income) or a 20% rate (if it is not).

The facts in this case raise the question whether the income that is shifted to Ireland is shifted from the United States or from the countries where the customers are located (the source or market countries). There is no doubt that some income is shifted from the market countries, but it is reasonably clear that the largest part of the value in Apple's products arises from its proprietary technology. Some is attributable to Apple's marketing, for which Apple U.S. makes a small charge to affiliates. It is doubtful that the preponderance of the Irish income is properly allocable to the in-country selling activity. In sum, for its non-U.S. sales Apple's use of cost sharing transfers the return to R&D performed in the United States to Ireland (or the ocean).

The tax motivation of Apple's income shifting is evident. The appropriate way to test the reality of the Apple arrangement is to ask whether Apple would have entered into this cost sharing arrangement if Apple's Irish affiliates had been unrelated. Over the three year period, 2009 – 2011 Apple's Irish cost sharing participants paid approximately $3.3 billion in cost sharing payments to Apple US. While that is a very large number, over the same period Apple's Irish affiliates earned EBT (after those payments) of $29.3 billion.[29] In other words, the $3.3 billion investment earned the right the substantial portion of $32.6 billion, or almost a 10 times return. The U.S. tax deferred likely is over $10 billion. The ability to reinvest those tax savings is a valuable tax benefit.

So, would Apple have entered into this cost sharing arrangement if Apple's Irish affiliates had been unrelated? To answer "yes" strains credulity.

The objective of the arm's length principle in transfer pricing is to achieve neutral treatment of related party and unrelated party transactions. The ability of multinational businesses to take advantage of transfer pricing between related persons in different countries strongly favors structuring transactions with affiliates to be able to shift income into low-taxed jurisdictions. It is an advantage that is largely unavailable to purely domestic businesses including most all small business enterprises. Yet, small businesses and individuals must make up the lost taxes.

There does not appear to be meaningful information regarding the effect of recently finalized cost sharing regulations on cost sharing. Anecdotally, it appears that companies have sought to grandfather existing agreements, as Apple has done, and are looking for other strategies for new projects.[30] This will bear monitoring closely. Of one point there is assurance, taxpayers will continue to focus on transfer pricing so long as there is potential to take advantage of material income tax differentials.

There are many potential steps that may and should be taken to improve the law and administration in respect of transfer pricing. I will discuss one proposal that transcends transfer pricing and bears consideration by the Subcommittee. There is a substantial need for more transparency by large public and comparable private companies. To date, companies do not routinely disclose information from consolidating financial statements with respect to the material separate legal entities of the consolidated group. Consolidating financial statements, which are unaudited separate company statements, are routinely prepared in connection with preparing an audited consolidated financial statement. These consolidating statements should be made available on a company web site with respect to each material company (with eliminations) along with information regarding the tax residency of each material company. This would provide valuable information to investors and analysts, who could monitor the group's assets and profitability by company, and more approximately by jurisdiction, and better assess the company's country and tax risks. This increased transparency would improve the monitoring of multinational businesses by shareholders, civil society and tax authorities alike and put downward pressure on corporate agency costs.[31]

Deduction Dumping

The benefit of income shifting is enhanced when deductions are incurred in the United States to earn low tax foreign income that is deferred from U.S. tax. Borrowing a table from Professor Harvey, below, it appears that Apple's general and administrative and sales, marketing and distribution expenses are incurred disproportionately in the United States. This helps explain the lower ratio of U.S. EBT to U.S. sales.

Allowing a current deduction for whatever portion of these expenses is attributable to income booked in the Irish companies (instead of in the United States) effectively is a U.S. tax subsidy those deferred earnings. Allowing the expense as a deduction, unreduced by the foreign earnings to which it is attributable (applying existing U.S. allocation rules), provides a tax saving benefit equal to the difference between the U.S. and foreign rate and the ability to invest that saving until the foreign earnings are distributed.

	Pre-tax Income/Sales	% of Pre-tax Income	General & Administrative Expenses		Sales, Marketing, & Distribution Expenses	
			$billions	%	$billions	%
US	24%	30%	1.7	85%	3.3	59%
Non-US	36%	70%	0.3	15%	2.3	41%
Consolidated	32%	100%	2	100%	5.6	100%

The allocation of deductions issue is a large dollar issue not only for Apple, but for the U.S. tax system more generally. In FY 2008, deductions allocable to foreign income (but not allocable to specific types of income) on Forms 1118 totaled $201 billion, including $99 billion of interest, $78 billion of other deductions (such as overhead expense) and $23 billion of R&D. The portion of these deductions properly allocable to deferred earnings should not be allowed as deductions until the deferred income is repatriated to the United States.

This issue would become even more significant if the United States were to shift to a dividend exemption for active foreign income.[32]

Sidestepping Anti-Deferral Rules

Deferral, and even more, exemption of foreign profits, creates an irresistible incentive to shift income to where it will be low-taxed or not taxed. This was understood when the Subpart F limits on deferral were first adopted in 1962 – they were intended to serve as a vital backstop against transfer pricing abuse by reducing the incentives that could arise if income could be shifted to low- or zero-tax countries. Apple's international structure takes full advantage of loopholes in existing anti-deferral rules. These rules have been substantially eroded, most significantly by ill-conceived application of "check-the-box" disregarded entity regulations in the international area. This problem was exacerbated by Congressional actions restricting a response to the problem. Additional exceptions that undermine the overall structure of Subpart F include an unprincipled expansion of the manufacturing exception to foreign base company sales income to cover contract manufacturing, the Section 954(c)(6) look-through rule and a "same country exception" based on place of incorporation. Apple avoids the reach of the foreign base company sales rules by contracting for manufacture of its products by third parties and in most cases, for U.S. tax purposes, selling to third parties. By using check-the-box disregarded entities, intercompany transactions within the group of companies that are classified as disregarded entities simply disappear.[33] With respect to payments of interest and dividends, the look-through rule of Section 954(c)(6) accomplishes much the same result except to the extent that deductible payments offset income of the payor that would be subject to current U.S. tax.

Finally, the same country exceptions for dividends and interest apply based on place of incorporation, whether or not the corporation is tax resident in the country of incorporation. Even before check-the-box and the look-through rule, taxpayers were taking advantage of nonresident Irish companies to sidestep this rule. If changes are made to check-the-box and look-through rules, changes also should be made to this same country exception. As a general proposition, if it is retained in anything like its present form, Subpart F should operate on a branch-by-branch basis and not by reference to place of incorporation.[34]

Implications of Apple Case Study - Where to Go From Here

Our international tax rules are out of balance. They are too generous to foreign income and not strong enough in protecting against U.S. base erosion by foreign companies investing in the United States. The losers are domestic business.

In the context of current law, changes may be made that would limit the scope for profit shifting. Most promising is a "minimum tax" imposed on the U.S. shareholder of a controlled foreign corporation in respect of low-tax foreign income earned by the controlled foreign corporation. In design, it actually would be a deemed distribution, as under current Subpart F, but the remaining U.S. tax would be collected when the earnings are distributed or the stock is sold. This approach would effectively take away the advantage of tax havens.

This should be accompanied by taking away the advantage of tax havens for foreign companies that invest in the United States. The United States should protect its source tax base by measures that may include imposing withholding tax on and/or restricting deductions for deductible payments of income paid to or treated as beneficially owned by related persons not "effectively taxed" on the income. In doing this, the United States would take away a substantial advantage that foreign-owned companies have in structuring investments in the United States.

Adopting a balanced approach is necessary to assure a level playing field. I have described elsewhere an approach that if taken by the United States would provide an incentive for other countries to adopt complementary rules. Moreover, the United States should strongly support and lead efforts at the OECD to combat base erosion and profit shifting. I acknowledge that the ideas described above need development into specific proposals, but this may be done in a reasonable time frame and will have value in relation to the principal international tax reform proposals.

Should Congress wait for tax reform to address income shifting? The short answer is "no." The two tax writing committees have begun work on a fundamental revision of the tax code. Many options on specific issues have been floated and a number of actual proposals put in draft legislative language. Some are good and some are bad. Like Vladimir and Estragon asking what Godot looks like, however, the players in the tax reform effort do not know what tax reform looks like. Without a coherent direction to the effort, including agreement on basic objectives and consistency in revenue estimating, the undertaking will founder or result in a messy patchwork of unstable political compromises. The political difficulty of the undertaking requires leadership from the Administration (centered in the Treasury Department, not the White House) as well as from the Hill. The technical complexity of the undertaking requires utilizing the knowledge and economic analysis skills of the Treasury Office of Tax Policy as well as the Staff of the Joint Committee on Taxation. The work on tax reform is at very early stages and will take years. Do not be lulled into "waiting for tax reform."

CONCLUSION

The Subcommittee is to be applauded for exposing international tax practices that are not easily discernible from public financial statements. The Apple case study adds further support to the findings from aggregate data that there is substantial shifting of profits offshore by U.S. multinationals.[35] Apple's income shifting strategies, including its cost sharing transfers of valuable intellectual property rights, are not unusual as evidenced in the 2010 case studies developed by the staff of the Joint Committee on Taxation and in the testimony presented in hearings by the U.K. Public Accounts Committee.[36] I encourage the Subcommittee to pursue reforms in the short term to adequately protect the U.S. tax base.

Thank you and I would be pleased to answer any questions.

End Notes

[1] My testimony is at the request of the Subcommittee, by letter dated May 1, 2013 from Chairman Carl Levin and Ranking Member John McCain. I am testifying in a personal capacity. My testimony does not represent the views of Harvard Law School or Harvard University.

[2] Prior to my current position, I was the Deputy Assistant Secretary for International Tax Affairs at the Department of the Treasury. Before my most recent government service, I was a tax partner at Ropes & Gray LLP for 22 years specializing in U.S. international income taxation before resigning in 2009 to serve in government. I occasionally consult for Ropes & Gray LLP on mutually agreed projects. I have provided a copy of my biography to the Subcommittee and a disclosure of my outside activities is posted on my faculty website page. Members of my family own Apple stock.

[3] The background portion of my testimony draws from my September 21, 2012, testimony before this Subcommittee. Readers familiar with these areas of law may wish to skip this background discussion.

[4] I.R.C. §§61(a)(7). The highest corporate tax rate is 35% for net income over $10 million. I.R.C. §11(b). The recapture of lower-bracket rates causes the corporate marginal rate to exceed 35% over limited income ranges. Earnings of a controlled foreign corporation may be deemed included in a United States shareholder's income under certain anti-deferral rules discussed below. See I.R.C. §§951 - 964.

[5] See I.R.C. §§901, 902, 904. The credit allowed for foreign income taxes is subject to a limitation, The credit for foreign income tax may not exceed the pre-credit U.S. tax that

otherwise would be paid by the taxpayer on foreign source net income in the same limitation category as the foreign tax. Today, there generally are two foreign tax credit limitation categories, one for passive income and another "general" category that includes all non-passive income. U.S. multinational taxpayers that earn high-tax foreign income, or that through planning "bunch" foreign taxes into high-tax pools of earnings used to repatriate foreign taxes for use as credits, may use excess foreign tax credits against other low-taxed foreign income in the same category. For example, excess foreign tax credits can be used to offset U.S. tax on royalty income and income from sales that pass title to customers outside the United States that is treated as foreign-source income for U.S. tax purposes (though this income generally would not be taxed by another country). See J. Clifton Fleming, Robert J. Peroni & Stephen E. Shay, Reform and Simplification of the U.S. Foreign Tax Credit Rules, 101 TAX NOTES 103 (2003), 31 TAX NOTES INT'L 1145 (2003).

[6] Subpart F is in Subchapter N of Chapter 1 of the Code. A controlled foreign corporation is a foreign corporation that is more than 50% owned, by vote or value, directly or indirectly under constructive ownership rules, by United States shareholders. I.R.C. § 957(b). A United States shareholder is a U.S. person that owns ten percent or more by vote, directly or indirectly under constructive ownership rules, of the foreign corporation. I.R.C. § 951(b). Passive income defined as "foreign personal holding income" in Code section 954(c) is one category of "foreign base company income" that is taxed currently.

[7] I.R.C. §§ 954(d) and 954(b)(4).

[8] I.R.C. §§ 954(d) and (e).

[9] Subpart F also has a de minimis exception if a controlled foreign corporation's foreign base company income is less than the lesser of 5% of gross income or $1 million and a "full inclusion" rule if more than 70% of a foreign corporation's gross income is foreign base company income. The discussion in the text is a summary of the relevant provisions and is not intended to be comprehensive. For example, the discussion does not cover, inter alia, the active foreign finance or insurance exceptions to Subpart F or foreign base company oil income.

[10] 2006 IRS Statistic of Income (SOI) data show that 12.2% of foreign earnings and profits of controlled foreign corporations (with positive current year earnings) were taxed currently under Subpart F. Statistics of Income, Table 3. U.S. Corporations and Their Controlled Foreign Corporations: Number, Assets, Receipts, Earnings, Taxes, Distributions, and Subpart F Income, by Selected Country of Incorporation and Industrial Sector of Controlled Foreign Corporation, Tax Year 2006, at http://www.irs.gov/taxstats/bustaxstats/article/0,,id=96282,00.html. An additional 7.9% of foreign earnings were distributed in a taxable distribution. Lee Mahony and Randy Miller, Controlled Foreign Corporations, 2006, STATISTICS OF INCOME BULLETIN 197, 202 Figure C (Winter 2011) (taxable payout ratio of 9.7% in relation to positive current year earnings and profits net of Subpart F income) see http://www.irs.gov/pub/irs-soi/11coforeign06winbull.pdf. When the 9.7% is measured in relation to positive current year earnings it is 7.2% (9.7% multiplied times the ratio of positive current year earnings and profits net of Subpart F income/positive current year earnings and profits (400,854,698/491,235,961) = 7.9%).

[11] Statistics of Income, Table 3. U.S. Corporations and Their Controlled Foreign Corporations: Number, Assets, Receipts, Earnings, Taxes, Distributions, and Subpart F Income, by Selected Country of Incorporation and Industrial Sector of Controlled Foreign Corporation, Tax Year 2006, at http://www.irs.gov/taxstats/bustaxstats/article/0,,id=96282,00.html and author's calculations. The average effective rate disguises far lower effective rates for

certain industries and companies, such as Apple. Companies in the resource industries often pay much higher levels of foreign tax.

[12] I.R.C. § 956. The rules were strengthened in the 1970s after a U.S. shipping magnate circumvented this restriction by using his controlled foreign corporation shares as collateral for a loan. Ludwig v. Comm'r, 68 T.C. 979 (1977), nonacq., 1978-2 C.B. 1. In response, regulations were amended with addition of a rule known to all U.S. multinational financing lawyers (and auditors) – a pledge of stock will be deemed to be an investment in U.S. property by the controlled foreign corporation if "at least 66 2/3rds percent of the total combined voting power of all classes of stock entitled to vote is pledged and if the pledge is accompanied by one or more negative covenants or similar restrictions on the shareholder effectively limiting the corporation's discretion with respect to the disposition of assets or the incurrence of liabilities other than in the ordinary course of business." Treas. Reg. §1.956-2(c)(2) (T.D. 7712, 1980). See Gustafson, Peroni & Pugh, TAXATION OF INTERNATIONAL TRANSACTIONS [¶6200- 6220] (4th Ed. 2011).

[13] I.R.C. §956(c).

[14] Accordingly, it is commonplace for a controlled foreign corporation to hold U.S. dollar bank deposits, U.S. government and corporate debt securities of unrelated issuers, and U.S. equity securities of unrelated issuers. A 2011 survey by the U.S. Senate Permanent Investigations Subcommittee majority staff estimated that of $538 billion of undistributed accumulated foreign earnings (of 27 surveyed multinationals as of the end of FY 2010) approximately 46% was invested in U.S. bank accounts and securities. U.S. Senate Permanent Investigations Subcommittee Majority Staff, Report Addendum to Repatriating Offshore Funds: 2004 Tax Windfall for Select Multinationals (Dec. 14, 2011).

[15] The benefit of deferral is not eliminated when the deferred earnings are reinvested in investments producing Subpart F income even when there is no U.S. interest deduction for the group. See generally, Myron S. Scholes, Mark A. Wolfson, Merle Erickson, Edward Maydew, Terry Shevlin, TAXES AND BUSINESS STRATEGY: A PLANNING APPROACH, 347-348 (4th Ed. 2009).

[16] I.R.C. §482; Treas. Reg. §1.482-1(a).

[17] See Seagate Technology, Inc. v. Comm'r, 102 T.C. 149 (1994); Veritas Software corp. v. Comm'r 133 T.C. 297 (2009).

[18] See I.R.C. §482 (second sentence).

[19] 1988-2 C.B. 458.

[20] U.S. Department of the Treasury, "Report to the Congress on Earnings Stripping, Transfer Pricing and U.S. Income Tax Treaties," (Nov. 2007).

[21] T.D. 9568, 76 FR 80082 (Dec. 22, 2011).

[22] Treas. Reg. §§1.482-1(e), 1.482-7.

[23] See J. Clifton Fleming, Jr., Robert J. Peroni and Stephen E. Shay, Worse than Exemption, 59 Emory LAW J. 79, 119-127 (2009).

[24] To preserve confidentiality, information only was made available at the Subcommittee offices or in the presence of a Subcommittee staff member. In the future, I suggest that the Subcommittee employ a secure virtual data site, which is customary practice in commercial merger, acquisition and financing transactions to preserve confidential company data.

[25] I refer to Apple's fiscal year ending September 24, 2011 instead of the most recently ended fiscal year because separate subsidiary information only was made available to the Subcommittee staff for FY 2011. The Apple consolidated numbers are from Apple's Form 10-K for FY 2011.

[26] This seems a reasonable adjustment in light of Apple's decision to issue $17 billion in debt to help finance a $55 billion stock buyback rather than repatriate earnings and reportedly pay $9.2 billion in tax. Peter Burrows, Apple Avoids $9.2 Billion in Taxes With Debt Deal, Bloomberg.com (May 3, 2013), at http://www.bloomberg.com/news/2013-05-02/apple-avoids-9-2-billion-in-taxes-with-debt-deal.html?cmpid=yhoo (last visited May 19, 2013). See also, Martin Sullivan, "Economic Analysis: Apple Reports High Rate But Saves Billions On Taxes," 2012 TNT 29-2 (Feb. 9, 2012).

[27] In 2011, the distribution of personnel and functions among Irish companies was somewhat mixed up and was rationalized in 2012. See description in Staff Memorandum. For purposes of describing numbers of employees and earnings before tax (EBT), I will treat the entities as one entity.

[28] The better measure for transfer pricing analysis is operating income, however, I use EBT for comparability reasons. Use of operating income would not affect the findings.

[29] As noted above, the better measure is operating income, but the numbers would remain enormous.

[30] It has been suggested that transferring existing intangible property in tax-free transactions so as to be subject to Section 367(d) rules avoids the reach of some of the rules of the cost sharing regulations. That certainly should not be correct in that Section 367(d) should not have a different outcome than Section 482.

[31] It should be possible to adopt standards that would address trade secret concerns. There is no public policy interest in basing market competition on transfer pricing and tax strategies.

[32] Proposals to use a 5% "haircut" in a possible U.S. dividend exemption system as a surrogate for allocating expenses materially understate the amount of deductions allocable to foreign income.

[33] It remains necessary to consider the application of the foreign base company sales rules for sales and manufacturing branches, but they also are fairly readily controlled.

[34] See American Bar Association Tax Section Task Force on International Tax Reform, "Report of the Task Force on International Tax Reform," 59 TAX LAW. 649, 787-809 (2006).

[35] In 2010, Treasury testimony reviewed a range of studies that indicate substantial income shifting to lower tax countries, including evidence from company tax data of margin increases correlated inversely with effective tax rates. The key conclusion of that review of studies based on aggregate data was that there was evidence of substantial income shifting through transfer pricing. Testimony of Stephen E. Shay, Deputy Assistant Secretary International Tax Affairs, U.S. Department of Treasury, House Ways and Means Committee, Hearing on Transfer Pricing Issues (July 22, 2010), http://democrats.ways andmeans.house.gov/media/pdf/111/2010Jul22_Shay_Testimony.pdf.

[36] See, Staff of Joint Committee on Taxation, Present Law And Background Related To Possible Income Shifting And Transfer Pricing, (JCX 37-10 2010);House of Commons, Committee of Public Accounts, HM Revenue & Customs: Annual Report and Accounts 2011–12, Nineteenth Report of Session 2012–13, 7- 12, Ev 21 – Ev 50 (HC 716, Dec. 3, 2012), at http://www.publications.parliament.uk/pa/cm201213/cmselect/cmpubacc/716/716.pdf (last visited March 16, 2013) (Oral Evidence Taken from Troy Alstead, Starbucks Global Chief Financial Officer, Andrew Cecil, Director, Public Policy, Amazon, and Matt Brittin, Google Vice President for Sales and Operations, Northern and Central Europe, on Monday, November 12, 2012).

In: Offshore Profit Shifting ... ISBN: 978-1-62808-479-5
Editor: Reny Toupin © 2013 Nova Science Publishers, Inc.

Chapter 5

TESTIMONY OF MARK J. MAZUR, ASSISTANT SECRETARY FOR TAX POLICY, U.S. DEPARTMENT OF THE TREASURY. HEARING ON "OFFSHORE PROFIT SHIFTING AND THE U.S. TAX CODE – PART 2 (APPLE INC.)"*

Chairman Levin, Ranking Member McCain, and members of the Subcommittee, I appreciate the opportunity to testify on the issue of the potential shifting of profits offshore and between foreign countries by U.S. multinational corporations.

This is a multifaceted, complex subject that raises numerous tax policy issues as well as issues relating to tax administration and tax accounting. My testimony, however, will be limited to tax policy considerations.

POTENTIAL SHIFTING OF PROFITS OFFSHORE BY U.S. MULTINATIONAL CORPORATIONS

The geographic allocation of profits earned by multinational enterprises historically has been challenging and has become more difficult with the rise of globalization. To see the complexity, consider a stylized example:

---

* This is an edited, reformatted and augmented version of a Testimony Presented May 21, 2013 before the Senate Permanent Subcommittee on Investigations.

- Employees at a U.S.-based firm come up with an idea for a new software application;
- They collaborate with a team of software engineers at a subsidiary in Country A to elaborate on the concept and develop the initial prototype;
- Employees at a subsidiary in Country B develop and test the Beta version and pilot it to a limited audience;
- Employees at a subsidiary in Country C modify the Beta version for commercial use;
- Software is distributed in the U.S., Europe, and Asia through company-owned cloud computing centers; while
- Employees at a subsidiary in Country D oversee all the contractual arrangements between the parties and also account for all the transactions between related and unrelated parties.

The question that arises is where the income from this product is earned. Presumably, some sliver of income should be attributed to each of the subsidiaries, but because all the steps were required to successfully market the product, the appropriate geographic allocation between the U.S. parent and each of the subsidiaries is not obvious.

However, the Internal Revenue Code ("Code") requires that income be allocated to the various subsidiaries based on the "arm's length" standard, which is essentially what unrelated parties would charge each other for the goods or services provided. But, when parties are related and where there is not a well-defined market, it may be problematic to determine the arm's length prices that should prevail on these transactions. And with more cross-border transactions taking place between related parties, this issue has become bigger over the last few decades. It is important to realize that this is not just a U.S. problem. Virtually every country with a corporate income tax faces the challenge of determining what share of a global enterprise's income is part of that country's tax base. Pushing in the other direction are trends in tax planning and accounting where multinational enterprises are creating what some commentators have called "stateless income," not subject to tax in the jurisdictions where the company is located and where it does business.[1]

Multinational corporations are able under current law to shift profits offshore and between subsidiaries located in different countries using various organizational structures and transactions. In some cases, a U.S. company transfers rights to intangible property to an offshore affiliate. Such cross-border transfers of intangible property rights could occur in various contexts, including cost-sharing arrangements. Under a cost-sharing agreement, a U.S. multinational corporation enters into an agreement with one of its controlled foreign corporations ("CFCs"), typically in a low-tax jurisdiction, in which both companies agree to share the costs and benefits of the development of intangible property. The CFC is required to pay the U.S. parent an arm's length amount for any existing intangible property or other resources it makes available for use in the shared research and development activities. Thereafter, the CFC contributes a percentage of the costs corresponding to its anticipated benefits from the intangible property to be developed (e.g., from the rights to exploit the intangible property in the CFC's territory). Under established transfer pricing principles, because the CFC bears its share of development costs, the CFC is entitled to the returns from exploiting the intangible property in its territory, which, in some instances, may be significant. This may be the case even if the CFC employs few people and otherwise performs few functions beyond the cost contribution and acting as owner of the intangible property.

In theory, the upfront arm's length payment for the intangible property originally contributed by the parent (reflecting the value of the property transferred), combined with the reduction in the parent's U.S. tax deductions, should result in no anticipated risk-adjusted loss of tax revenue to the U.S. as compared to the case in which no cost-sharing agreement is entered into. However, there has been considerable controversy about whether this result is achieved in fact.

Further, some other U.S. tax rules (e.g., the "check-the-box" rules and the Subpart F CFC look-through rule) allow U.S.-based multinationals to redeploy profits earned by the CFC from exploiting the intangible property to related CFCs (or other customers/licensees) without incurring a U.S. level of income tax. Under U.S. tax rules, the profits of foreign corporations are not subject to U.S. income tax until the profits are repatriated to U.S. persons, unless the profits constitute Subpart F income (discussed below). The postponement of taxation until repatriation is commonly referred to as deferral.

In other transactions, profits of foreign subsidiaries may be shifted by assigning certain risks to a minimal-activity foreign affiliate in a lower-tax jurisdiction. Such an affiliate may be treated as a "principal" earning profit (in the form of a risk premium) with respect to ongoing activities that continue to be conducted by the "de-risked" transferor.

Additional ways that U.S. multinationals may shift profits include moving intangible property (and related profits) offshore through various transactions that may not result in recognized income for U.S. tax purposes. In general, transfers of intangible property by a U.S. person to a non-U.S. corporation would result in a deemed royalty to the U.S. transferor under Code Section 367(d) over the useful life of the property that is commensurate with the transferee's income from the property. However, taxpayers sometimes take the position that this outcome does not apply to certain intangibles (such as workforce in place). In addition, taxpayers sometimes take the position that a disproportionate amount of intangible value represents foreign goodwill and going concern value (i.e., the value of a corporation to potential buyers as a continuing operation), which are explicitly carved out of the Section 367(d) regulations. Similarly, taxpayers sometimes take the position that foreign goodwill, going concern value, and workforce in place are not covered by the current definition of intangible property in the Code, so that their transfer is not subject to the arm's length transfer pricing rules of Code Section 482.

CHANGES IN U.S. CORPORATE INCOME TAX RATES

Changes in U.S. corporate income rates – both in absolute terms and relative to the rates of our major trading partners – have changed the economic incentive for the shifting of profits. Before 1987, the U.S. maximum statutory corporate income tax rate was relatively high (between 46 percent and 53 percent from the 1950s through 1986) and roughly similar to those of other industrialized countries. The 1986 Tax Reform Act reduced U.S. income tax rates and broadened tax bases significantly and the maximum statutory corporate rate has remained at 34 percent or 35 percent since. Through the late 1990s, the U.S. corporate tax rate tended to be below the average for developed countries but since then, due to reductions in foreign corporate income tax rates, it has been above average and is now among the highest in the developed world.

A higher statutory rate can encourage companies to shift income and production to a lower-tax jurisdiction, especially in today's global marketplace. The immediate financial gain from shifting a dollar of income from one jurisdiction to another equals the difference in statutory income tax rates between the two locations. And while there may be costs to managing operations and earnings that have been shifted between jurisdictions, the multinational firm may still be better off from having done so. In addition, the statutory corporate income tax rate may also affect the decision to invest in one country rather than another, especially where the investments are independent and highly profitable.

ACCOUNTING TREATMENT OF DEFERRED EARNINGS

U.S. multinationals are concerned not just about the tax treatment of their earnings but also about the financial accounting treatment. There is a presumption under U.S. Generally Accepted Accounting Principles (GAAP) that deferred income taxes should be recognized in the financial statements for the same period in which the earnings are generated because U.S. GAAP presumes that the foreign earnings will be remitted to the U.S.-based parent company at some point in time in order to distribute the earnings to shareholders. This presumption may be overcome if the firm develops sufficient evidence that the foreign entity has permanently invested or will permanently invest the earnings in the foreign jurisdiction. Accordingly, the deferral of earnings offshore not only offers a tax benefit (lower effective tax rate paid in the current accounting period) but may result in higher earnings for financial statement purposes (by presuming that the U.S. corporate income tax will never be paid on these "permanently" reinvested earnings). Thus, financial income reporting rules may add to the incentive to shift earnings.

REVENUE LOSS FROM PROFIT SHIFTING

Estimates of the potential revenue loss to the U.S. government from profit shifting cover a wide range, from $10 - $20 billion to well over $80 billion per year. These estimates attempt to consider profit shifting from all sources, including non-arm's length transfer pricing on intercompany trade with affiliates, strategic location of debt, and transfers and location decisions involving intangibles. One prominent estimate showed a revenue loss to the Federal government of $87 billion for 2002. This estimate included shifting by both U.S.-based multinationals and foreign-based multinationals operating in the United States [2] Other estimates are lower, for example, one indicated a revenue loss of $17.4 billion for U.S.-based companies in 2004, though this only attempted to measure the additional shifting that occurred in 2004 relative to that which occurred in 1999 and not total shifting.[3]

Author	Annual Tax Shift	Year of Estimate
Clausing (2009)	$87 b	2002
Christian and Shultz (2005)	$30 b	2001
Sullivan (2008)	$17.4 b	2004

A different way to develop estimates of the magnitude of profit shifting involves estimating the potential revenue gain from adopting specific policies intended to restrict income-shifting opportunities. In this regard, the Joint Committee of Taxation estimated that the revenue gain from completely repealing deferral would be around $11 billion in 2010.

One study of sales-based formulary apportionment approaches to allocating income to geographic locales estimates that its adoption would have raised $50 billion in 2004.[4] However, that estimate does not incorporate all the behavioral responses by companies if formulary apportionment were implemented. A different analysis that attempts to simulate various behavioral responses concludes that formulary apportionment would raise no more revenue than the current system.[5]

All these estimates are necessarily based on a set of assumptions about behavior and profitability. For example, some studies assume that rates of return or profit margins in the United States and foreign locations would be the same if there were no income shifting. Others try to estimate statistical relationships between profitability in a country and tax differentials. And most of these studies are based on financial data published by the Commerce Department, not tax data. That said, these studies provide some insight into the potential magnitudes of profit shifting and the effect on Federal revenues.

OVERVIEW AND HISTORY OF THE SUBPART F RULES

The Subpart F rules attempt to prevent the shifting of income, either from the United States or from the foreign country in which it was earned, into a low- or no-tax jurisdiction. Thus, Subpart F generally targets both passive and mobile income. The Subpart F rules discourage the shifting of these types of income by disallowing deferral of U.S. taxation for such income and requiring current taxation. (In related party transactions, the shifting of income may be achieved more easily because a commonly controlled group of corporations can direct the flow of income between entities in different jurisdictions.)

The Subpart F rules are set forth in Code Sections 951-964 and apply to certain income of CFCs. The Code defines a CFC as a foreign corporation more than 50 percent of which, by vote or value, is owned by U.S. persons, each of whom owns a 10 percent or greater interest in the corporation by vote (each a "U.S. shareholder"). The term "U.S. persons" includes U.S. citizens or residents, domestic corporations, domestic partnerships, and domestic trusts and estates. If a CFC has Subpart F income, each U.S. shareholder must include its pro-rata share of that income in its gross income as a deemed dividend in the year the income was earned. Thus, this income is taxed at the U.S. tax rate in the year earned (that is, the tax on this income is not "deferred").

Subpart F was enacted in 1962 during the Kennedy Administration. Key rationales for its enactment included preventing tax abuse, taxing passive income currently, promoting equity, promoting economic efficiency, and avoiding undue harm to the competitiveness of U.S. multinationals. While the Kennedy Administration proposal would have ended deferral for all income earned by foreign subsidiaries of U.S. taxpayers, Congress was concerned that ending deferral completely would place U.S. companies at a competitive disadvantage in their foreign operations. The enactment of Subpart F was a more modest step toward ending deferral, focused on the types of income that were viewed as more easily shifted.

The Subpart F rules have been modified since 1962. For example, in 1976, a new foreign base company shipping income category was added.[6] In 1982, a new category of foreign oil-related income was added. In 1986, many changes were made to the Subpart F rules, including the expansion of the foreign personal holding company income category to include income from commodities (unless derived in the active conduct of a qualifying commodities business), gains from the disposition of many types of property and certain foreign currency gains. In 1997, the foreign personal holding company income category was expanded to include income from notional principal contracts and substitute dividend payments. Additionally, a temporary exception for income derived from the active conduct of a banking, financing or insurance business was added after being removed 11 years earlier (and this exception has been extended multiple times, most recently in the American Taxpayer Relief Act of 2012). A look-through rule providing an exception to foreign personal holding company income for payment of dividends, interest, rents and royalties out of active earnings was added in 2006.[7]

Subpart F may, in some cases, not be doing what it was intended to do. It is possible for taxpayers to avoid some important provisions of Subpart F, due in part to the proliferation of hybrid entities and hybrid instruments. Hybrid entities are entities that are classified as flow-through entities in one jurisdiction (for example, the United States) and as corporations in another jurisdiction. Hybrid instruments are financial instruments that are treated as debt in one jurisdiction and as equity in another jurisdiction. Therefore, it is now possible in some cases to shift income to low- or no-tax jurisdictions and earn passive income in such jurisdictions without triggering Subpart F and having this income taxed in the U.S. as it is earned.

THE CHECK-THE-BOX RULES

Several observers have noted that the proliferation of techniques involving hybrid entities has lessened the effectiveness of the current Subpart F regime. Although not the exclusive source of these planning techniques, the "check-the-box" entity classification regulations, which became effective January 1, 1997, have resulted in significantly increased use of hybrid entities. And while initially not aimed at foreign affiliates, these rules have been substantially used by multinational firms.

The availability of tax avoidance techniques involving hybrid entities did not originate with the checkthe-box regulations. However, the check-the-box regulations exacerbated the problem in three significant ways. First, they eliminated the uncertainty associated with applying the existing test for entity classification. This reduced the costs and risks associated with hybrid arrangements and thus greatly facilitated their use. Second, they focused attention on the use of hybrid arrangements. The result was a considerable increase in design and marketing efforts among tax planners that introduced hybrid planning techniques to mainstream taxpayers. Finally, and perhaps most importantly, the check-the-box regulations facilitated the formation of a new type of entity (or non-entity): an entity "disregarded as an entity separate from its owner" (often referred to as a "disregarded entity"). The disregarded entity features prominently in a number of Subpart F tax planning techniques.

The Administration is concerned about the misuse of various income-shifting devices, including misuse of the check-the-box rules, to inappropriately avoid the Subpart F rules, and thus has proposed legislative changes to tighten rules and reduce incentives that encourage the shifting of investment and income overseas.

SECTION 954(C)(6) LOOK-THROUGH RULE

Congress enacted the so-called "look-through rules" under Section 954(c)(6) as part of the Tax Increase Prevention and Reconciliation Act of 2005. Section 954(c)(6) allows one CFC to make payments of dividends, interest, rents, and/or royalties to a related CFC without resulting in Subpart F income to the recipient CFC so long as the amounts are attributable to income of the payor CFC that is neither non-Subpart F income nor income effectively connected with the conduct of a U.S. trade or business. Section 954(c)(6) was intended to allow U.S. multinational corporations to redeploy their active foreign earnings without incurring a level of U.S. tax.

Section 954(c)(6) was enacted as a temporary provision effective for taxable years beginning after December 31, 2005, and before January 1, 2009. Since then, Section 954(c)(6) has been extended three times (most recently in the American Taxpayer Relief Act of 2012) and is currently in effect through taxable years beginning before January 1, 2014.

IMPACT OF THE CHECK-THE-BOX RULES AND SECTION 954(C)(6)

The check-the-box rules and Section 954(c)(6) both result in a higher amount of earnings being eligible for deferral. Deferral encourages U.S. multinationals to keep earnings offshore.

The Treasury Department estimates that the U.S. revenue impact of these provisions is on the order of a few billion dollars per year, mainly because the provisions reduce the after-tax cost of foreign activity and therefore encourage such activity. The provisions also reduce repatriation of profits to the parent company, albeit with a higher U.S. residual tax rate for the funds that are repatriated. Absent these provisions, the shifting of profits from high- to low-tax foreign countries would occur less frequently and would incur greater costs. The United States generally would not directly receive significant additional revenue as a result of the profits not being shifted, but U.S. multinationals would pay higher foreign taxes through their foreign subsidiaries (and thus to the extent these earnings were repatriated, there would be a lower residual U.S. tax, after foreign tax credits).

ADMINISTRATION INITIATIVES TO REDUCE THE SHIFTING OF PROFITS OFFSHORE

The President's Framework for Business Tax Reform is intended to strengthen the international tax system. The proposals for reform take a multipronged approach that reduces incentives for companies to shift profits and investment to low-tax countries, puts the United States on a more level playing field with our international competitors, and helps slow (or perhaps end) the global race to the bottom on corporate tax rates. There is considerable debate as to how to reform the international tax system, but there appears to be common ground on this subject, including a shared concern about preserving the U.S. tax base by reducing incentives for the shifting of investment and income overseas and about making the United States a more attractive place to create and retain high-quality jobs.

The President's Framework would impose a minimum rate of tax on the income earned by the foreign subsidiaries of U.S. multinationals. This would discourage companies from moving profits offshore.

Foreign income otherwise subject to deferral in a low-tax jurisdiction would be subject to immediate taxation up to the minimum tax rate with a foreign tax credit allowed for income taxes on the income paid to the host jurisdiction. This minimum tax would be designed to provide a balance by limiting the opportunities to shift profits to lower-tax jurisdictions while also placing U.S. multinationals on a more level playing field with local competitors.

The President's Framework for Business Tax Reform also would incorporate many of the international tax proposals included in the President's FY 2014 Budget that would discourage U.S. multinationals from shifting profits (and specifically profits related to intangible property) offshore. Under one such proposal, a new category of Subpart F income would be added for excess profit returns from intangibles that have been transferred by a U.S. person to a related CFC. Specifically, this proposal provides that if a person transfers intangible property from the U.S. to a related foreign affiliate that is subject to a low foreign effective tax rate in circumstances evidencing excess income shifting, then the U.S. person must include in income currently the amount equal to the excessive return.

A second proposal would clarify the scope of intangible property that is subject to the deemed-royalty rules of Section 367(d) and the transfer pricing rules of Section 482 to include workforce in place, goodwill and going concern value. Another proposal addresses the concern that, under current law, a U.S. business can borrow money and invest overseas and take a current deduction for the interest related to overseas investment, even though the U.S. business may pay little or no U.S. taxes on the income from the overseas investment. The Administration's proposal would eliminate this tax advantage by requiring that the deduction for interest expense attributable to the overseas investment be matched with the income it is supporting (that is the deduction for interest expense would be delayed until the related income is taxed in the U.S.).

Furthermore, the Treasury Department and the Internal Revenue Service (IRS) have issued regulations and other guidance to discourage the shifting of profits offshore.

In 2008, the Treasury Department and the IRS issued comprehensive temporary regulations under Section 482 pertaining to cost-sharing arrangements.

These temporary regulations, which became effective on January 5, 2009, and were finalized in 2011, clarified a number of contentious issues and better defined the scope of intangible property transfers and contributions that require compensation.

Early anecdotal information indicates that the regulations have had a positive impact on taxpayers' reporting positions. As an important complement to the cost sharing regulations, in 2009, the Treasury Department and the IRS also finalized regulations covering service transactions, including services performed using high value intangibles.

Additionally, the Treasury Department and the IRS have recently issued regulations under Section 909 that limit the use of foreign tax credits in situations in which foreign taxes are inappropriately separated from (and taken into account in advance of) the underlying foreign income with respect to which the foreign taxes were paid. The regulations defer the ability to claim a foreign tax credit for foreign taxes until the related income is taxed in the United States. The Treasury Department and the IRS have also issued regulations under Section 367(a)(5) that make it more difficult to move earnings offshore in tax-free reorganization transactions, and Notice 2012-39 reduced incentives to move intangible property offshore as part of a tax-free repatriation strategy.

Finally, the Treasury Department supports the efforts of the Organisation for Economic Co-Operation and Development (OECD) to analyze these profit-shifting issues and is actively participating in the OECD's projects to address these issues, including the project analyzing Base Erosion and Profit Shifting. This important multilateral effort is evidence that governments around the world are wrestling with these difficult issues and trying to find ways to address inappropriate profit-shifting.

Thank you, and I look forward to answering your questions.

End Notes

[1] See, e.g., Edward D. Kleinbard, The Lessons of Stateless Income, 65 TAX L. REV. 99 (2011); Edward D. Kleinbard, Throw Territorial Taxation from the Train, 114 TAX NOTES 547 (Feb. 5, 2007).

[2] Clausing, Kimberly A., "Multinational Firm Tax Avoidance and Tax Policy," National Tax Journal, Vol LXII, No. 4, December 2009, pp. 703-724. This estimate attributes the difference in profitability between U.S. multinational firms and their affiliates abroad to differences in the U.S. and host country tax rates and allocates the profit difference to the United States based on the share of affiliate transactions that occur with the United States relative to the share that occurs with affiliates in other countries. This approach does not take into account the myriad other factors that may affect differences in profitability.

[3] Sullivan, Martin A., "U.S. Multinationals Shifting Profits Out of the United States, Tax Notes, March 10, 2008, pp. 1078-1082.

[4] Clausing, Kimberly A. and Reuven Avi-Yonah, Reforming Corporate Taxation in a Global Economy: A Proposal to Adopt Formulary Apportionment, Brookings Institution: The Hamilton Project, Discussion paper 2007-2008, June 2007.

[5] Altshuler, Rosanne and Harry Grubert, "Formula Apportionment: Is it Better Than the Current System and Are There Better Alternatives," National Tax Journal, forthcoming.

[6] The foreign base company shipping income category was repealed in 2004.

[7] For more information about the development of the Subpart F rules, see Treasury's Policy Study: "The Deferral of Income Earned through U.S. Controlled Foreign Corporations" (December 2000).

In: Offshore Profit Shifting …
Editor: Reny Toupin

ISBN: 978-1-62808-479-5
© 2013 Nova Science Publishers, Inc.

Chapter 6

TESTIMONY OF SAMUEL M. MARUCA, DIRECTOR, TRANSFER PRICING OPERATIONS, INTERNAL REVENUE SERVICE. HEARING ON "OFFSHORE PROFIT SHIFTING AND THE U.S. TAX CODE – PART 2 (APPLE INC.)'*

INTRODUCTION

Chairman Levin, Ranking Member McCain, and members of the Subcommittee, thank you for the opportunity to testify on tax compliance and administration issues related to the shifting of profits offshore by U.S. multinational corporations.

The IRS takes very seriously the need to ensure that U.S. multinational corporations are abiding by U.S. tax laws and paying their fair share of tax. Over the last few years, we have been working to enhance our approach to international tax enforcement in general and to offshore profit shifting in particular. We have been refocusing our enforcement efforts to be more strategic by viewing taxpayers through the prism of their tax planning strategies and allocating our limited resources to cases presenting the highest compliance risk.

---

* This is an edited, reformatted and augmented version of a Testimony Presented May 21, 2013 before the Senate Permanent Subcommittee on Investigations.

In implementing this new approach, we began from the premise that we need to determine where companies are using legitimate strategies to manage global tax exposure and where they may be pushing the envelope too far. Thus, we have been aligning our resources and training our employees in key strategic areas such as income shifting, deferral planning, foreign tax credit management, and accessing profits accumulated offshore through repatriation transactions.

To better manage our collective knowledge in strategic international compliance areas, we have formed 18 International Practice Networks, which are focused on integrating our training and data management with our strategy. We have also established a new International Practice Service, which will serve as a central repository for the knowledge and expertise of our international staff. For example, in the income shifting area, an international practice network is in the process of developing 25 different training and job aid tools, and over 400 international staff members have been participating in regular network calls devoted to income shifting topics.

As the IRS works to address tax avoidance issues involving multinationals, it is also important that we continue to work with other countries. At the multilateral level, the IRS and the Treasury Department are active participants in the Organisation for Economic Co-Operation and Development, where we are currently participating in several major guidance projects. The goal is to develop a coordinated and comprehensive action plan to update international tax rules to reflect modern business practices while preventing inappropriate cross-border profit-shifting.

CURRENT ISSUES IN TAXATION OF U.S. MULTINATIONALS

The IRS' enforcement authority in regard to profit shifting by U.S. multinational corporations arises primarily from section 482 of the Internal Revenue Code, under which the IRS is charged with ensuring that taxpayers report the results of transactions between related parties as if those transactions had occurred between unrelated parties. Under the section 482 regulations, as well as under multinational transfer pricing guidelines, the determination of whether the pricing of a transaction reflects an arm's length result is generally evaluated under the so-called "comparability standard." Under this standard, the results of the transaction as reported by the taxpayer are compared to results that would occur between by unrelated taxpayers in comparable transactions under comparable circumstances.

Establishing an appropriate arm's length price by reference to comparable transactions is relatively straightforward for the vast majority of cross-border transactions involving transfers of goods or services. But enforcing the arm's length standard becomes much more difficult in situations in which a U.S. company shifts to an offshore affiliate the rights to intangible property that is at the very heart of its business – what may be referred to as the company's "core intangibles." In fact, over the past decade, applying section 482 in these types of cases has been the IRS' most significant international enforcement challenge.

When the rights to the core intangibles of a business are shifted offshore, enforcement of the arm's length standard is challenging for two reasons:

- First, transfers of a company's core intangibles outside of a corporate group rarely occur in the market, so comparable transactions are difficult, if not impossible, to find. In some cases the IRS has had to resort to other valuation methods, which are often referred to as "income-based methods." Under these types of methods, the IRS typically has to conduct an ex ante discounted cash flow analysis. This means that we are required to evaluate the projections of anticipated cash flows that the taxpayer used in setting its intercompany price; then we must further evaluate how the taxpayer discounted those projected cash flows to compensate for the risk associated with earning them. The challenge here is that evaluating the underlying assumptions made by the taxpayer, without benefit of hindsight, is not an exact science.

- Second, a business's core intangible property rights are by their nature very "risky" assets. So projecting cash flows from these assets and the appropriate discount rate requires an inherently challenging assessment of the underlying risk and how, and by which party, that risk is borne. These can be difficult assessments to make, at least in some cases.

Outbound international tax planning involves not only shifting profits to low-tax jurisdictions, but also managing exposure to the Code's anti-deferral provisions under subpart F. Subpart F requires U.S. shareholders of controlled foreign corporations (CFCs) to include currently in income for U.S. tax purposes their pro rata share of certain of the CFC's income – including dividends, interest, rents, royalties, and income from certain sales and services transactions. However, because subpart F contains many exceptions, careful planning allows companies to avoid subpart F inclusions and even to enhance income shifting to low-tax jurisdictions.

Commonly, a company's strategy involves the making of deductible payments from foreign affiliates operating in high-tax jurisdictions to affiliates organized in low-tax jurisdictions. For example, if a low-tax affiliate lends to a high-tax affiliate, the interest expense related to that loan offsets the higher taxes imposed on the affiliate paying the interest, and the interest income received by the recipient affiliate is subject to a low, or zero, rate of tax. Under the original framework of the subpart F regime, the interest or royalty income received by the low-tax affiliate would constitute subpart F income and therefore would be taxable to the U.S. parent of the multinational group. Taxpayers, however, have long been able currently to avoid subpart F through various techniques.

For example, avoidance of subpart F on foreign-to-foreign deductible payments was facilitated with the issuance of the check-the-box regulations in 1997. Under these regulations, an eligible business entity can elect its classification for federal tax purposes. Of particular note, the check-the-box regulations provide that an eligible foreign entity with a single owner can be treated as "disregarded" as a separate entity and therefore taxed as a branch for U.S. purposes. As a result, deductible payments – such as interest and royalties – paid between the disregarded entity and its owner (or between two disregarded entities with the same owner) are ignored for U.S. tax purposes and avoid subpart F treatment. Importantly, these payments continue to be regarded for foreign tax purposes and thus reduce taxable income in the high-tax foreign jurisdiction.

Today, taxpayers can also rely on the so-called "CFC look-through rule" under section 954(c)(6) to avoid subpart F treatment on deductible payments without resorting to the check-the-box regulations. This rule excludes from subpart F income dividends, interest, rents, and royalties paid by one foreign affiliate to another affiliate, to the extent the payment is out of non-subpart F earnings of the payor.

Once profit is shifted to a low-tax foreign affiliate, and subpart F is avoided, U.S. multinationals will seek to repatriate offshore cash to the United States with minimal tax consequences. Simply dividending the cash to a U.S. affiliate will result in U.S. taxation at a 35-percent rate, reduced by a credit for any foreign tax imposed on the earnings. So U.S. multinationals seek ways to repatriate cash through sophisticated structures they assert do not result in dividend treatment. This is another area in which we are dedicating enforcement resources to ensure that these transactions are treated appropriately.

IRS ACTIONS TO IMPROVE
TAX COMPLIANCE
BY MULTINATIONALS

Transfer Pricing

The IRS' approach to the income shifting challenge is evolving. In the early 2000s, the IRS formed teams of experts known as issue management teams, or IMTs, to focus on transfer pricing and related business practices. These teams were made up of IRS transfer pricing specialists and Chief Counsel attorneys, led by IRS industry executives, and centrally managed the "inventory" of examinations involving transactions in these respective areas. The teams ensured that IRS resources were appropriately dedicated to these examinations, that best practices and processes were shared, and that the IRS position on the underlying issues was applied uniformly to cases under similar facts and circumstances.

In 2011, a new IRS executive position was created to oversee all transfer pricing-related functions, to set an overall strategy in the area, and to coordinate work on our most important cases. Further, in building a new function devoted exclusively to tackling our transfer pricing challenges, we recruited dozens of transfer pricing experts and economists with substantial private sector experience to help us stay on the cutting edge of enforcement and issue resolution.

Transfer Pricing Operations is divided into two parts. First is the Transfer Pricing Practice, which collaborates with other international personnel and industry groups to identify strategic work in the transfer pricing area and ensure appropriate development and presentation of cases with strategic merit. Second is the Advanced Pricing and Mutual Agreement program (APMA), which was created a year ago through the merger of our Mutual Agreement and Advanced Pricing Agreement programs. These new functions operate as a unified team with a global focus, a unified strategy, and a robust knowledge base.

Cost Sharing

The IRS has worked with the Treasury Department over the last several years to adopt revised regulations on cost sharing. In 2008, new section 482 regulations pertaining to cost sharing transactions were issued. These temporary regulations were effective on January 5, 2009, and were finalized in 2011. They clarify a number of issues that had been contentious under the previous set of cost sharing regulations and better define the scope of intangible property contributions that are subject to taxation in connection with cross-border business restructurings. While to date the IRS has had limited experience in auditing transactions covered by the new cost sharing regulations, early anecdotal information indicates that the regulations have had a positive impact on taxpayers' reporting positions in the area.

However, concerns remain that we are considering and following closely. Some taxpayers are taking the position that a cost sharing arrangement, or other transaction taxable under section 482, has been preceded, either explicitly or implicitly, by an incorporation or reorganization transfer of core intangibles. In these cases, the taxpayers assert, among other positions, that foreign goodwill and going concern value are the most valuable elements in these transfers. In response, we are now training our agents to address these issues and to challenge taxpayers' positions where appropriate.

Repatriation of Earnings

Focusing on the repatriation area, Treasury and the IRS over the past six years have issued several anti-abuse notices – one as recently as July 2012 – making clear that a variety of transaction types give rise to inappropriate repatriation results. In several of these cases, Treasury and the IRS have already followed up with regulatory changes necessary to make clear what the appropriate results should be.

In general, these transactions were designed to take advantage of mechanical rules which are scattered through the Code and regulations, and which pertain to determinations of either tax basis or earnings and profits. These rules were not written with repatriation in mind, and the transactions in which the rules have been used may not look like repatriation transactions at first blush – so they can be difficult to find. But we are finding them and where we have, we have acted quickly.

As to specific repatriation strategies being challenged by the IRS, these often involve foreign affiliates entering into various transactions with their U.S. parent that result in the parent receiving cash, notes or other property from the affiliates. Taxpayers assert that these transactions do not result in a dividend or gain to the U.S. parent corporation under various corporate non-recognition provisions. Examples of these transactions include so-called "Killer B" transactions, "Deadly D" transactions, zero-basis structures, and outbound F reorganizations. While these types of transactions have been addressed by new regulations, for pre-effective date periods the IRS has challenged many of them under common law doctrines and will continue to do so.

Taxpayers have also attempted to avoid dividend treatment by manipulating the amount and timing of a foreign subsidiary's earnings and profits. The IRS has challenged these types of transactions under existing law and has had some success. For example, in *Falkoff v .Commissioner*, the Seventh Circuit Court of Appeals reversed a Tax Court holding that a corporation's distribution in advance of recognizing earnings had economic substance.

Moreover, taxpayers may be able to offset residual U.S. tax on foreign earnings by using foreign tax credits. For example, taxpayers have implemented so-called "splitter" transactions to free up foreign tax credits for use to offset U.S. tax on repatriated low-taxed earnings. The IRS has challenged such transactions, under both the applicable provisions of the Code and underlying regulations and various judicial doctrines. Further, legislation enacted in 2010, *i.e.*, section 909, and the regulations published thereunder in 2012, should put a stop to many of these transactions.

Foreign corporations also enter into various repatriation transactions that are disguised loans to their U.S. parent corporation. Taxpayers assert that these transactions are not subject to section 956 and therefore do not result in income inclusion to the U.S. parent. The IRS has challenged, and will continue to challenge, these types of transactions under the applicable provisions of the Code and regulations, and under various judicial doctrines such as the doctrine of substance over form. For example, in *Merck & Co. Inc.* the Third Circuit Court of Appeals held that interest rate swaps entered into with foreign subsidiaries constituted a disguised loan subject to section 956.

Further, to address abusive short-term loan transactions like the one highlighted by this Subcommittee in the past, we developed and delivered specialized training for our employees on these issues. In April 2013, we conducted a three-hour online training session focusing on section 956, which was attended by more than 250 international examiners. This training session, which remains available online to all international employees, covers the general anti-deferral rules under section 956, as well as the exception for short-term loans, avoidance planning techniques, and audit techniques. We are also developing detailed job aid tools related to the section 956 short-term loan exception and the techniques being used to exploit it.

Casework: Examinations and Litigation

The IRS has been, and continues to be, vigilant and forceful in addressing compliance issues we have seen in regard to U.S. multinationals. Based on a recent survey, as of May 9, 2013, we estimate that we are currently considering income shifting issues associated with approximately 250 taxpayers involving approximately $68 billion in potential adjustments to income.

As for litigation in the income shifting area, the IRS has challenged approximately 34 transfer pricing issues involving 15 taxpayers in 22 U.S. Tax Court cases over the past three years. Of those 22 cases, the IRS litigated and lost two: *Xilinx v. Commissioner*, 125 T.C. 37 (2005), *aff'd*, 598 F.3d 1191 (9th Cir. 2010), and *Veritas v. Commissioner*, 133 T.C. No. 14 (2009). In *Xilinx*, the IRS included stock-based compensation as a cost to be shared in a cost sharing arrangement. Unfortunately, the court did not sustain the government's position. In *Veritas*, the IRS challenged the taxpayer's buy-in amount under the cost-sharing arrangement by applying an income method. In this case as well, the court rejected the government's approach and sustained the taxpayer's buy-in amount with some adjustments.

CONCLUSION

Mr. Chairman, Ranking Member McCain, thank you again for this opportunity to testify on the IRS' efforts to enforce the tax law as it applies to U.S. multinational corporations. Although enforcing and administering this section of the tax law will present challenges for the IRS into the future, the agency has made great strides in recent years, and this is a tribute to strategic focus and to the highly dedicated and professional men and women of the IRS. I would be happy to answer any questions you have.

INDEX

D

E

G

F

H

I

J

L

M